Digital Filters and Signal Processing in Electronic Engineering
Theory, Applications, Architecture, Code

"Talking of education, people have now a-days" (said he) "got a strange opinion that every thing should be taught by lectures. Now, I cannot see that lectures can do so much good as reading the books from which the lectures are taken. I know nothing that can be best taught by lectures, except where experiments are to be shewn. You may teach chymestry by lectures — You might teach making of shoes by lectures!"

James Boswell: *Life of Samuel Johnson, 1766* (1709-1784)

"Mathematics possesses not only truth, but supreme beauty – a beauty cold and austere like that of sculpture, and capable of stern perfection, such as only great art can show.

Bertrand Russell in *The Principles of Mathematics* (1872-1970)

About the Authors

S.M. Bozic graduated in communications and electronics at Belgrade University and subsequently worked in research and development in the communications industry, in Germany and the UK for 11 years. He obtained his PhD at the University of Aston (UK) and then joined the School of Electronic and Electrical Engineering at the University of Birmingham (UK), as a lecturer. His main areas of interest started in communications and electronics, but for some years, his research and teaching activity has specialised in discrete-time and digital circuits, leading to digital filters, digital control and particularly Kalman filtering. Dr. Bozic has published a number of research papers on digital systems, as well as books, including *Digital and Kalman filtering*, (Arnold 1979, 1994), *Electronic and switching circuits* (co-author, Arnold 1975) He also contributed to *Digital signal processing*, (IEE Control Engineering, University of Sussex 1982).

R.J.Chance graduated firstly in medical biochemistry, with a further degree in electronics engineering at the University of Birmingham, UK. He has specialized in microprocessors since 1976 and joined the Microprocessor Systems Laboratory at Birmingham University when it was formed in 1980. As manager of this facility, his remit was to assist all departments in Birmingham University to introduce and use microprocessors in their teaching and research activities. In the course of this, he was responsible for the design of more than 40 microprocessor-based systems, applied to just about everything between biochemistry and transportation. During 1982, Jim Chance started to specialise in digital signal processors, particularly the design of development environments for DSP system testing. Around 30 published papers describe this and other work. The TMS320 assembler and simulator used in this book were written by RJC. In 1990, he joined the School of Electronic and Electrical Engineering at Birmingham as a lecturer, where he has been responsible for second and third year courses on microprocessors, as well as industrial research and development in collaboration with the railway traction, nautical navigation and telecommunications industries. R.J.Chance is a member of the British Computer Society.

Digital Filters and Signal Processing in Electronic Engineering

Theory, Applications, Architecture, Code

S.M. BOZIC

and

R.J. CHANCE
School of Electrical and Electronic Engineering
University of Birmingham

Horwood Publishing
Chichester

First published in 1998 by
HORWOOD PUBLISHING LIMITED
International Publishers
Coll House, Westergate, Chichester, West Sussex, PO20 6QL
England

11765143

Learning Resource
Centre

British Library Cataloguing in Publication Data
A catalogue record of this book is available from the British Library

ISBN 1-898563-58-6

Printed in Great Britain by Martins Printing Group, Bodmin, Cornwall

Preface

The term Digital Signal Processing (DSP) is normally used to describe signal processing in discrete time, leading to difference equations for the case considered. However, the execution of these equations as algorithms on a Digital Signal Processor is also called Digital Signal Processing, even though the expertise demanded of the engineer is **utterly different from that required for the mathematical basis**. Until recently, such implementations were mostly simulations of a process described by an algorithm. However, present day microprocessors of specialised design - DSP devices- are capable of performing DSP operations in real time.

A special feature of this book is that, after deriving the equations, they are implemented as algorithms, not only in Matlab, but, through pseudo-code, right up to working DSP-based software whose speed and performance can be directly demonstrated. We anticipate that this book will give the reader the confidence to perform the same type of development process to suit his own particular work. The construction of state-of the-art engineering products invariably involves the design of working machinery, based on fundamentals which can only be expressed mathematically. However, at the sharp end of the research and development business, one needs to have a physical model of the working system in mind at all times, in order to be able to make modifications in the field with confidence. This is why the reader will find **unique non-mathematical conceptual models** of several important but perplexing basic ideas, such as the FFT and multirate filters, to complement the mathematical theory.

Chapter 1 introduces the general signal processing scheme, from the sampling process to fundamental results in the form of the Discrete-time Fourier Transform (DtFT). The reader proceeds through DSP signal descriptions, such as **difference equations, transfer functions and convolution to understand concepts of multirate DSP**.

Chapter 2 Describes how numbers are stored and manipulated within the digital signal processor, which is so often quite different from methods used in conventional computers. This material is **independent of any manufacturer's processor and applicable to any DSP device**.

Chapter 3 gives a brief insight into the architectures of 8 DSP devices from four manufacturers in order to demonstrate special hardware features which are architecture dependent. This should put the reader in a good position to **use and make a comparison between different current and future DSP devices**.

Chapter 4 develops the DFT and its use in convolution and correlation and a rapid method for carrying out DFT computations, the **Fast Fourier Transform (FFT).** **The Goertzel algorithm and Chirp-z transform** are introduced. All these start as mathematical concepts and in Chapter 4 are **carried through to programs running on the Matlab system**.

Chapter 5 **explains the FFT,** which is so very well known for its ability to separate a waveform into its component sine-waves, **as a physical model**. This can be used to "intuit" the effects of all sorts of manipulations of the algorithm, in support of the mathematical rigor of chapter 4. Assembler language programs for both FFT and Goertzel are constructed from first principles and their performance given.

Chapter 6 Explains the **properties and design of Finite Impulse Response (FIR) filters** by Fourier series and frequency response sampling, with examples including a multirate filter design. Algorithms are demonstrated, using Matlab and a multirate filter scheme is carried through to the DSP assembler language stage. This includes a non-mathematical construction of **coefficient ordering and other multirate filter principles**.

Chapter 7 covers **IIR filter structures** and design by impulse invariance and bilinear transform methods. There is also consideration of the effects of **quantization noise and scaling**. Some of the effects of the latter on practical design can be seen in an assembler language IIR example, constructed, as usual from the equations, via Matlab and pseudocode.

Chapter 8 concludes the book with some other important DSP topics. These include deconvolution and system identification, homomorphic deconvolution, control and adaptive filtering. **The reader is taken through the whole software development process for a control system and an adaptive filter, whose performance is shown**.

Very few individuals are able to master the whole of the progression from equation to working device, in any field, but especially if it is as new as digital signal processing. The intention of this book is to remedy this difficulty for up-and-coming young engineers, working with all manner of digital filters, using Fourier, thinking about applying specialised DSP devices, which can be applied in so many engineering areas.

We hope that students using this book will no longer feel that they need to create programs by slavishly copying existing listings and attempting to adapt them to the immediate purpose without full comprehension. We hope that the student will understand and build on the examples between these covers, at all stages, from mathematical foundation to operational machine. This might become a lifelong and profitable habit.

S.M. Bozic and R.J.Chance

Birmingham, UK,
September 1998.

ACADEMIC REQUIREMENTS

First and second year maths and introductory knowledge of microprocessors and the C language.

INTERNET SOFTWARE

The TMS320C25 assembler, simulator and DSP source listings in this book can be found on the internet website HTTP://www.eee.bham.ac.uk/dspbook.

Contents

1

Basic concepts and analytical tools

1.0   INTRODUCTION

The general digital signal processing scheme is shown in Fig. 1.1(a). The central element is the digital signal processing (DSP) unit preceded at the input side by the analogue to digital converter (ADC) and followed by a digital to analogue converter (DAC). The role of the lowpass filters (LPF) at each end will be explained in the following section. The scheme shown refers to analogue signal processing in which case ADC and DAC units are necessary. If the signal to be processed is already in digital format, and the result of the processing is to be used in digital form, then ADC and DAC would be omitted.

The ADC unit is expanded in Fig. 1.1(b) to show its constituent parts. The sampler is followed by the quantizer (with 2^b quantizing levels), and coder producing binary sequences in groups of b-bits. The DAC unit Fig. 1.1(c) consists of the decoder followed by the sample and hold (S/H) unit, producing y(n) from [y_b(n)] set of binary values.

This chapter consists of six sections. It starts with sampler analysis which produces fundamental results relating time and frequency domains of sampled signals. The second section deals with quantization noise introduced by the ADC unit. The following two Sections, 1.3 and 1.4, introduce the z-transform and inverse z-transform which are important analytical tools for the sampled data systems generally referred to as the discrete-time systems. The fifth section presents different ways of DSP unit signal description, as indicated in Fig 1.1(d). The final Section 1.6, introduces concepts of multirate digital signal processing which is used in many applications.

The DSP unit digital structure and its operating features are described in Chapter 2.

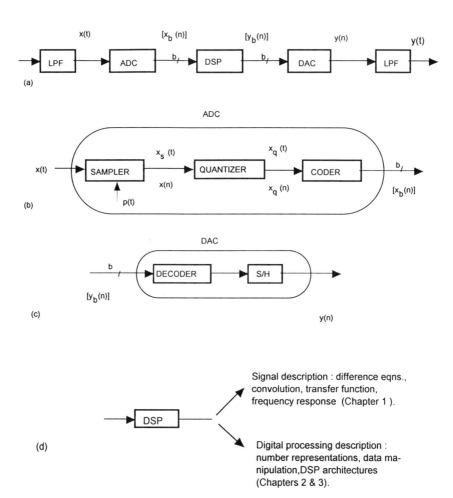

Fig. 1.1: General digital signal processing scheme

1.1   SAMPLING AND SIGNAL RECONSTRUCTION

In this section we relate the frequency spectrum of the continuous-time (analogue) waveform to its sampled version. The spectrum of the continuous-time waveform is assumed to be band limited to ω_m, Fig. 1.2(a), where the triangular shape is arbitrary. The input LPF in Fig.1.1, often referred to an antialiasing filter, is used to ensure that the bandwidth of the signal to be sampled is limited to ω_m. It also limits the additive noise and other interfering signals.

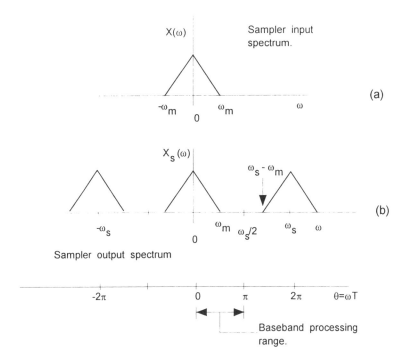

Fig. 1.2: Sampler input and output spectrum

The sampling process in Fig. 1.1(b) can be modelled by the following multiplication operation

$$x_s(t) = x(t) \cdot p(t)$$

where $\quad p(t) = \sum_{n=-\infty}^{\infty} \delta(t - nT)$ (1.1)

is an infinite sequence of Dirac delta functions at T seconds apart (Appendix 1.3). The spectrum of the sampled waveform $x_s(t)$, derived in Appendix 1.1, is given by

$$X_s(\omega) = \sum_{n=-\infty}^{\infty} x(nT) \, e^{-jn\omega T} \quad , \qquad\qquad (1.2)$$

where $x(nT)$ is a sample of $x(t)$ at the discrete-time nT. On the other hand, as shown in Appendix 1.2, the sampled waveform spectrum can be also expressed as

$$X_s(\omega) = \frac{1}{T} \sum_{r=-\infty}^{\infty} X(\omega + r\omega_s) \quad . \qquad\qquad (1.3)$$

This result shows that $X_s(\omega)$ is a periodic function consisting of $X(\omega)$ and shifted replicas of $X(\omega)$ at sampling frequency intervals $\omega_s = 2\pi / T$, Fig. 1.2(b).

By inspection of $X_s(\omega)$ we note that to avoid spectral interference (known also as aliasing) in the baseband range, it is necessary that $\omega_s - \omega_m \geq \omega_m$, hence $\omega_s \geq 2\omega_m$. This determines the minimum sampling frequency, for this idealised band-limited case, to be $\omega_s = 2\omega_m$ (sampling theorem); in many practical cases $\omega_s > 2\omega_m$.

        Fig. 1.2(b) also shows that to obtain the continuous-time signal, y(t), from its samples y(n) one needs to use the lowpass filter (LPF) as indicated in Fig. 1.1(a). This filter, also called a reconstruction filter, ensures that only the baseband range appears at the output of the processing system.

        The Fourier series pair, Eq. (A1.4) is periodic in time with the repetition period T. Here, we have $X_s(\omega)$, Eq. (1.2), which is periodic in frequency with the repetition period ω_s, so by analogy with Eq. (A1.4), we can write the corresponding second equation

$$x(nT) = \frac{1}{2\pi} \int_{-\pi/T}^{\pi/T} X_s(\omega)\, e^{jn\omega T}\, d\,(\omega T)\ , \tag{1.4}$$

where we have used $\omega_s = 2\pi/T$. This is a suitable place to introduce new variables $\theta = \omega T$, and $X(\theta)$ instead of $X_s(\omega)$, and rewrite Eqs. (1.2) and (1.4) as follows:

$$X(\theta) = \sum_{n=-\infty}^{\infty} x(n)\, e^{-jn\theta}\ , \tag{1.5}$$

$$x(n) = \frac{1}{2\pi} \int_{-\pi}^{\pi} X(\theta)\, e^{jn\theta}\, d\theta\ . \tag{1.6}$$

        These equations describe the discrete-time Fourier transform (DtFT). We shall be referring to them later in other chapters.

1.2   ADC QUANTIZATION NOISE

An ADC performs sampling, quantization and coding of the input signal, Fig. 1.1(b). The input x(t) is the analogue signal, x(n) is an exact sample of x(t) at time nT and $x_q(n)$ is x(n) rounded to the nearest quantization level. The quantization step size (or the resolution of the ADC) is given by

$$q = \frac{R}{2^{b+1}}\ , \tag{1.7}$$

where R is the peak-to-peak range of the quantizer, and (b + 1) bits is the word length.

        The error due to quantization, $e(n) = x_q(n) - x(n)$, is assumed to be uncorrelated with x(n) or with itself, and it is uniformly distributed over

[-q/2, q/2]. The mean value of the error is zero, and the variance (the quantization noise power) is [23,27]

$$\sigma_e^2 = \frac{q^2}{12} \quad . \tag{1.8}$$

The signal to quantization noise power ratio (SQNR) is given by

$$SQNR = 10 \ \log \ \frac{\sigma_x^2}{\sigma_e^2} \quad dB \ , \tag{1.9}$$

where σ_x^2 is the signal power. Using Eqs. (1.7) and (1.8) in (1.9) we have

$$SQNR = 6\,b + 16.8 - 20 \ \log \frac{R}{\sigma_x} \quad dB \quad . \tag{1.10}$$

Various expressions for SQNR are obtained depending on the type of signal applied to ADC. For example if the input is a sinusoidal signal, then $R/\sigma_x = 2\sqrt{2}$ and

$$SQNR = 6\,b + 7.9 \quad dB \quad . \tag{1.11}$$

For a random input signal with Gaussian probability distribution, the probability of $|x(n)| > 4\sigma_x$ is very low. So, setting the range of the quantizer to $R = 8\sigma_x$, less than six out of every 10^5 input signal samples would result in an overload on the average. For $R = 8\sigma_x$, Eq. (1.10) gives

$$SQNR = 6\,b - 1.25 \quad dB \quad . \tag{1.12}$$

The expression in (1.10) is often used to specify the precision needed in an ADC. It shows that each additional bit in the quantizer increases the SQNR by 6 dB.

1.3   THE Z-TRANSFORM

This is an important tool for the analysis of discrete-time systems. It plays the same role in the analysis of discrete-time systems as the Laplace transform in the continuous-time systems.

We develop this transform from the Fourier transform of sampled sequence, derived in Appendix 1.1, and given here as

$$X_s(\omega) = \sum_{n=-\infty}^{\infty} x(nT)e^{-jn\omega T} \qquad . \qquad (1.13)$$

The factor $e^{-jn\omega T}$ is due to the time delay from the term $\delta(t-nT)$ in Eq. (A1.2). We now make the following changes to Eq. (1.13):

(i)   To cover the whole of the s-plane, we change $j\omega$ to s.

(ii)   Then, introduce the new variable

$$z = e^{sT} \qquad (1.14)$$

(iii)   With the sampling time T = 1, x(nT) becomes x(n).

(iv)   Change $X_s(\omega)$ to X(z).

With these changes Eq. (1.13) becomes

$$X(z) = \sum_{n=-\infty}^{\infty} x(n)z^{-n} \qquad , \qquad (1.15)$$

known as the two-sided z-transform of the sequence x(n). This transform requires that the signal samples be specified for the entire time range - $\infty < n < \infty$, which is not suitable for many practical situations with nonzero initial conditions. Therefore, we use the one-sided z-transform given by

$$Z[x(n)] = X(z) = \sum_{n=0}^{\infty} x(n) z^{-n} \qquad , \qquad (1.16)$$

where the lower limit is zero. This transform is unique only for signals starting from n=0.

The z-transform of a signal sequence delayed by m time units is given by

$$Z[x(n-m)] = \sum_{n=0}^{\infty} x(n-m)z^{-n} = z^{-m} X(z) \qquad (1.17)$$

for zero initial conditions, i.e. x(n - m) = 0 for n < m. For nonzero initial conditions see, for example, [27, Section 4.5]. Eqs. (1.16) and (1.17) will be applied later.

The mapping between s and z planes, given by Eq. (1.14) has a physical meaning which is illustrated in Fig. 1.3(a). Both variables are complex and

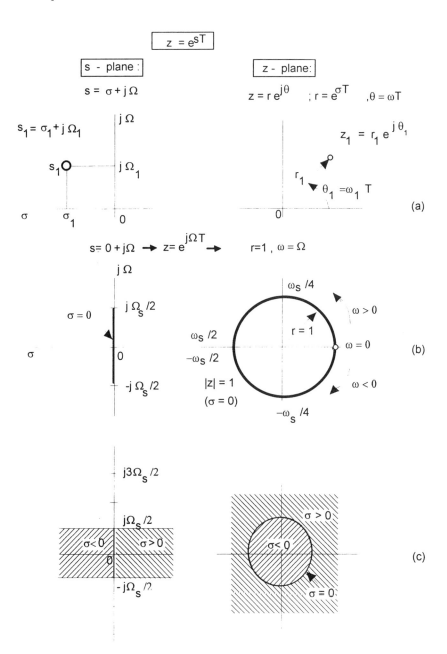

Fig. 1.3:

    (a) Mapping of point in s-plane to point in z-plane

    (b) Mapping of imaginary axis in s-plane to unit circle in z-plane

    (c) Mapping of s-plane areas into z-plane areas

expressed as $s = \sigma + j\omega$, and $z = r\,e^{j\theta}$. Then, from $z = e^{sT}$ we have $r\,e^{j\theta} = e^{\sigma T}\,e^{j\omega T}$ hence the real part of s determines the radius of the circle in z-plane, and the imaginary part (real frequency ω) determines the angle $\theta = \omega T$.

Part (b) in Fig.1.3 shows the special case ($\sigma = 0$) for which the frequency range $\pm\omega_s/2$ is mapped onto unity circle in z-plane. This is an important relationship to be met later in filter design.

Part (c), in Fig.1.3, shows mapping of areas in the s-plane into corresponding areas in the z-plane. Note that the stable area of the s-plane maps inside the unit circle of the z-plane.

$x(n), n \geq 0$		$X(z)$
Unit pulse:	$\delta(n)$	1
Unit step:	1	$1/(1-z^{-1})$
Exponential:	c^n	$1/(1-cz^{-1})$
Ramp:	n	$z^{-1}/(1-z^{-1})^2$
Cos-wave:	$a^{nT}\cos bnT$	$(1-az^{-1}\cos bT)/(1-2az^{-1}\cos bT + a^2 z^{-2})$
Sine-wave:	$a^{nT}\sin bnT$	$(az^{-1}\sin bT)/(1-2az^{-1}\cos bT + a^2 z^{-2})$

Table 1.1: Table of commonly used z-transforms

1.4   INVERSE Z-TRANSFORM

This means, given $X(z)$ we want to find the corresponding time sequence $\{x(n)\}$ which is formally expressed as

$$\{x(n)\} = Z^{-1}[X(z)] \quad .$$

Examples: The simplest case: $X(z)$ is a polynomial.
Given:

$$X(z) = 1 + 0.5z^{-1} + 0.25z^{-2} + 0.1z^{-3} \tag{1.18}$$

then $\quad\quad \{x(n)\} = \{1, 0.5, 0.25, 0.1\} \quad ,$

which is obtained directly from polynomial coefficients. However, $X(z)$ would be more often given as a rational polynomial, for example

$$X(z) = \frac{1}{1 - 0.2z^{-1} - 0.8z^{-2}} \qquad (1.19)$$

$$= \frac{z^2}{z^2 - 0.2z - 0.8} \cdot$$

The simplest way of obtaining $\{x(n)\}$ is by means of long division resulting in

$$\{x(n)\} = \{1, 0.2, 0.84, 0.33, 0.74, 0.41, 0.67,....\}.$$

This approach is practical only for obtaining approximate inversion giving a few terms. For a complete, or closed form solution one needs to use partial fractions, as illustrated below. In this case, we rewrite the given function, Eq.(1.19), as

$$X(z) = z\left[\frac{z}{(z-1)(z+0.8)}\right] , \qquad (1.20)$$

where the function within the square brackets has the degree in its numerator less than the degree of its denominator (proper rational function). Splitting the function within the square bracket into a sum of two fractions, and using them in Eq.(1.20), we obtain

$$X(z) = \frac{1}{1.8}\left(\frac{z}{z-1}\right) + \frac{1}{2.25}\left(\frac{z}{z+0.8}\right) . \qquad (1.21)$$

Comparing terms with the z-transform Table 1.1 we can see that the first term corresponds to the step function, the second to the exponential function.

$$\therefore x(n) = \frac{1}{1.8} + \frac{1}{2.25}\left(-0.8^n\right) , \qquad (1.22)$$

which produces sequence in agreement with the long division result.

A third method of inverting a z-transform $X(z)$ is based on the contour integration. For our purpose it is expressed as follows

$$x(n) = \frac{1}{j2\pi} \int_C X(z) z^{n-1} dz$$

$$= \sum_i \left[\text{residue of } X(z) z^{n-1} \text{ at } p_i \right] \qquad (1.23)$$

$$= \sum_i \left[(z - p_i) X(z) z^{n-1} \right]_{z=p_i} \qquad ,$$

provided that the poles p_i, are distinct (simple), and are inside the contour C.

For an illustration we use the case with X(z) given by Eq. (1.19) which has two poles $p_1 = 1$ and $p_2 = -0.8$. Applying (1.23) we have

$$x(n) = \left| (z - 1) \frac{z^2}{(z-1)(z+0.8)} z^{n-1} \right|_{z=1} + \left| (z + 0.8) \frac{z^2}{(z-1)(z+0.8)} z^{n-1} \right|_{z=-0.8} .$$

$$x(n) = \frac{1}{1.8} + \frac{1}{2.25}(-0.8)^n$$

This result agrees with (1.22), but it has been obtained without using a z-transform Table. The contour C is a circle of radius greater than one, with distinct (simple) poles inside C. With multiple-order poles one has to use a modified approach [16,23,27].

1.5   DSP SIGNAL DESCRIPTION

The discrete-time signals in Fig. 1.1, x(n) at the output of the sampler and y(n) at the output of the DAC unit, are related in a linear time-invariant system by the following **difference equation**

$$y(n) = \sum_{m=0}^{N} b_m x(n-m) - \sum_{m=1}^{N} a_m y(n-m) \qquad . \qquad (1.24)$$

The coefficients a_m and b_m are determined by the type of operation required by the application (filtering, signal generation, modulation etc.) To simplify without loss of generality, the upper limits have been taken to be the same (= N).

Applying the z-transform to Eq.(1.24) we have

$$Z[y(n)] = \sum_{m=0}^{N} b_m Z[x(n-m)] - \sum_{m=1}^{N} a_m Z[y(n-m)]$$

and using the relationships in Eqs.(1.16) and (1.17) we obtain

$$Y(z) = X(z)\left(\sum_{m=0}^{N} b_m z^{-m}\right) - Y(z)\left(\sum_{m=1}^{N} a_m z^{-m}\right) \qquad .$$

The above equation enables us to define **the transfer function** as the ratio of $Y(z)$ and $X(z)$,

$$H(z) = \frac{\displaystyle\sum_{m=0}^{N} b_m z^{-m}}{1 + \displaystyle\sum_{m=1}^{N} a_m z^{-m}} \qquad . \qquad (1.25)$$

This is an important result which will be used later. Note that for the nonzero initial condition $H(z)$ would have additional terms, [27, Section 4.6.2].

The **frequency response** is obtained directly from the above $H(z)$ by using the relationship in Eq.(1.14), and setting $s = j\omega$, which results in

$$H(\omega) = H\left(z = e^{j\omega T}\right) = \frac{\displaystyle\sum_{m=0}^{N} b_m e^{-j\omega mT}}{1 + \displaystyle\sum_{m=1}^{N} a_m e^{-j\omega mT}} \qquad . \qquad (1.26)$$

This is a periodic function as discussed in Section 1.1, but we are interested mainly in its shape within the baseband, i.e. $0 \le \omega \le \omega_s/2$, where ω_s is the sampling frequency in radians/s.

The zeros of $H(z)$, Eq. (1.25), are the values of z for which $H(z) = 0$. In Eq. (1.25) there are N zeros at $z_1, z_2, ..., z_N$ (roots of the numerator polynominal), and N poles at $p_1, p_2, ..., p_N$ (roots of the denominator polynomial). Therefore, the transfer function Eq. (1.25) can be expressed in terms of **poles and zeros** as

$$H(z) = G\frac{\displaystyle\prod_{i=1}^{N} (z - z_i)}{\displaystyle\prod_{i=1}^{N} (z - p_i)} \qquad , \qquad (1.27)$$

where G is a constant. This is a very useful representation for the analysis and design of discrete-time systems. Other important design features are obtained by setting the coefficients $a_m = 0$, in Eq. (1.25) which produces

$$H(z) = \sum_{m=0}^{N} b_m \, z^{-m} \quad .$$

(1.28)

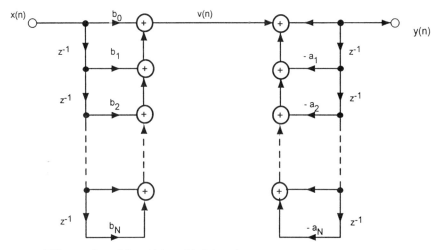

FIR: $a_m = 0$, m=1,2,...,N; $b_m = h(m)$: impulse response

$$y(n) = v(n) = \sum_{m=0}^{N} b_m \, x(n-m)$$

IIR: $a_m \neq 0$ $\qquad y(n) = v(n) - \sum_{m=1}^{N} a_m y(n-m)$

Fig. 1.4: Direct form implementation of Eq. (1.24)

This is an all-zero system from which **the convolution equation** is derived, in Appendix 1.4, as

$$y(n) = \sum_{m=0}^{N} h(m) \, x(n-m) \, ,$$

(1.29)

where $h(m) = b_m$ is the impulse (unit sample) response. For the general pole-zero transfer function, Eq. (1.25), the upper limit in the convolution equation becomes infinite (Appendix 1.4).

On this basis we can classify discrete-time systems into the finite impulse response (FIR) described by Eq. (1.28) or (1.29), and the infinite impulse response (IIR), Eq. (1.25). There are other names attached to this division. The FIR is also known as the moving-average (MA) model, while the IIR is the autoregressive-moving average (ARMA) model. The implementation of Eq. (1.24), known as the direct form 1, is shown in Fig. 1.4. The left part is the feedforward unit (FIR), and the right is the feedback unit (IIR). The FIR and IIR structures have different properties and different design methods to be seen later in Chapters 6 and 7.

More detail for the material in Sections 1.1 to 1.5 can be found in [16, 23, 27, 33].

1.6   MULTIRATE DIGITAL SIGNAL PROCESSING

Such processing uses more than one sampling rate to perform the desired digital system operation. The idea behind a multirate system is to decrease the sampling rate, carry out the required processing with a reduced number of operations, and then restore the original sampling rate. The process of decreasing the sampling rate is called decimation and that of increasing the sampling rate interpolation. As shown in Fig. 1.5, the sampling rate decrease is denoted by a factor D, and the increase by a factor I. By making D = I, the input and output sampling rates will be the same, and the system appears externally as a single rate system.

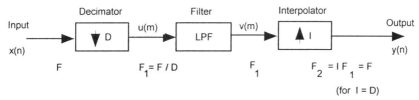

Fig. 1.5: Multirate filtering scheme

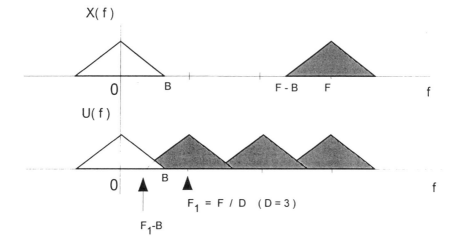

Fig. 1.6: Decimator spectrum

The multirate systems are time-varying and hence the transfer function and impulse response concepts developed for time-invariant systems do not apply.

There is well developed theory of multirate systems for further studies [9,33]. However, in the following subsections we develop the important concepts in decimation and interpolation using a simplified approach.

1.6.1   Decimation

The decimator consists of a sampling rate decreaser (SRD) which ignores (D-1) samples. This is easy to implement but care must be taken to avoid aliasing as indicated in Fig. 1.6. Aliasing can be prevented if the bandwidth of the original is such that (F/D) - B ≥ B. This can be achieved by preceding the SRD with a lowpass filter, Fig. 1.7, whose cut-off frequency should be less than or equal to F/2D. Such a filter can be considered as a digital anti-aliasing filter since it operates on digital signals. (The more familiar analogue anti-aliasing filter operates on an analogue signal prior to sampling).

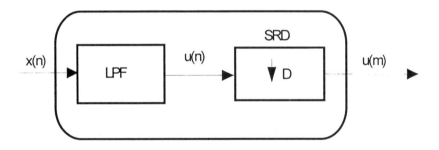

Fig. 1.7: Decimator structure

In principle the digital anti-aliasing filter can be realised as a standard FIR structure, Fig. 1.8(a), but this would be very inefficient. The filter would compute every new value of u(n) for a new x(n), yet the decimator ignores (D-1) samples out of every D samples. Great computational savings can be made if the filter is only made to calculate those values required at the output of the SRD. This can be achieved by putting the SRD before each multiplier as shown in Fig. 1.8(b). The filter calculations are now performed after decimation, at a rate F/D, resulting in a more efficient structure. The SRDs are identical and operate in step with each other, i.e. allow one sample each through. Note that m is given by the division in n/D.

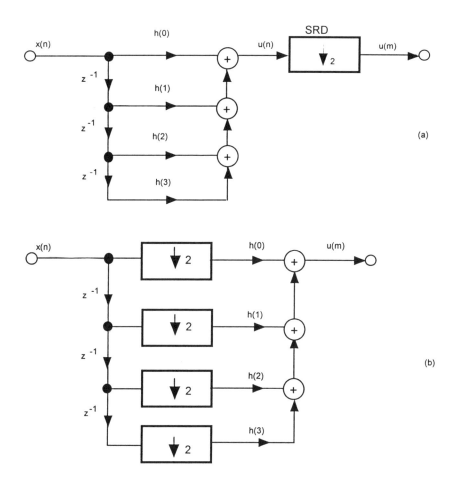

Fig. 1.8: FIR realisation of a decimator (D = 2, N = 4)

1.6.2   Interpolation

In the above subsection, it was shown that the decimator consists of a lowpass anti-aliasing filter followed by a sampling rate decreaser. In a similar way it is shown here that an interpolator can be considered as a sampling rate increaser (SRI) followed by a lowpass filter, Fig. 1.9.

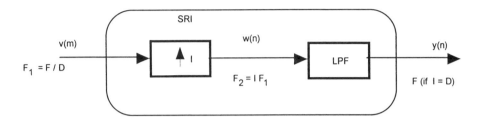

Fig. 1.9: Interpolator structure

The SRI inserts (I-1) zero samples between input samples. Its operation and the need for a lowpass filter is deduced from the frequency spectra in Fig. 1.10, for the points in Fig. 1.9, as follows. The input spectrum to the SRI, coming from the decimator with $F_1/2 = B$, Fig. 1.10(a), consists of the baseband centred at zero frequency, with the replicas at F_1, $2F_1$, $3F_1$,.... shown shaded. Assuming that the SRI operates at $F_2 = 3F_1$, its output spectrum $W(\omega)$ would be as shown in Fig. 1.10(b), the baseband now extends up to $F_2/2$, and the first repetition is centred at F_2. To obtain the original signal we have to filter out the band between $F_1/2$ and $F_2/2$, resulting in the spectrum shown in Fig. 1.10(c). Clearly, the lowpass filter needs to have a cut-off frequency of $F_1/2$. (Note this is a digital LPF and hence it has a periodic frequency response at the sampling frequency $3F_1$) This is called an anti-imaging filter and should be compared to the anti-aliasing filter used in decimation; they are identical for $I = D$.

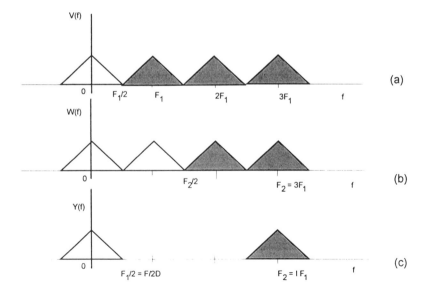

Fig. 1.10: Interpolator spectra

Again, as in the case of the anti-aliasing filter, we consider an FIR structure for the anti-imaging filter. Since the filter appears after the SRI unit, it is suitable to use a transposed FIR structure as shown in Fig. 1.11(a). Transposition: if all the branches in a signal-flow graph are reversed in direction, and the roles of the input and output are interchanged, the input/output response of the (transposed) network is unchanged [9].   The zero samples inserted by the SRI result in redundancy in the filter calculations. By putting the SRIs after the multipliers, a computationally more efficient structure results, Fig. 1.11(b), which operates at a lower rate than that in Fig. 1.11(a).

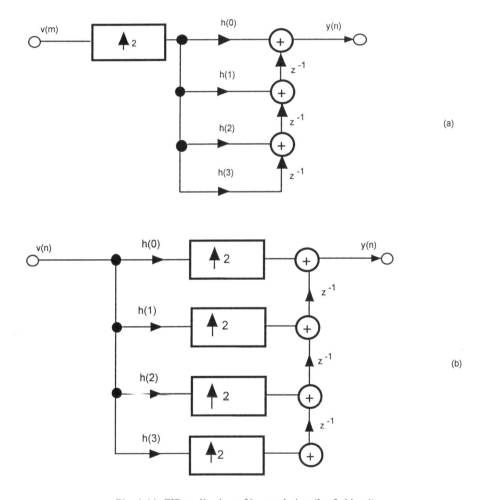

Fig. 1.11: FIR realisation of interpolation (I = 2, N = 4)

We have briefly discussed the FIR implementation of both the decimator and interpolator. This is the simplest realisation and it also provides linear phase response. The IIR implementations are also interesting and they can be found in [9,33].   There is another type of implementation by means of polyphase (or Npath) filter structures. These offer a computationally even more efficient solution. They have also been omitted here because they are conceptually more complex, but their description can be found in Section 6.5.3 and [9,27,33].

1.6.3   Combined decimation and interpolation

In the previous two subsections, the decimator and interpolator were examined separately. We now consider combinations of these units, starting with the cascade connection shown in Fig. 1.12(a). In this case the anti-imaging filter and the anti-aliasing filter follow each other directly. These filters operate at the same sampling rate (FI); but the one with higher cut-off frequency is redundant as indicated in Fig. 1.12(b). This structure can be used to change the sampling rate by any rational number I/D.

Consider next the cascaded structure with the decimator preceding the interpolator, Fig. 1.13. Both filters now operate at a lower sampling rate and have a cut-off frequency of F/2D. For the special case D = I, the structure in Fig. 1.13 would just filter the signal, while that in Fig.1.12 would leave it unaltered (just delayed). Therefore, Fig. 1.13 represents a filter with a cut-off frequency f_c = f/2D, and by suitable choice of D can be used to implement any lowpass filter specification.

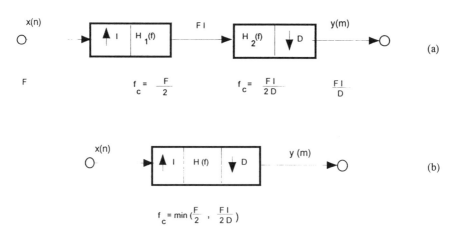

Fig. 1.12: Interpolator/Decimator

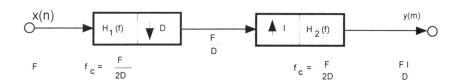

Fig. 1.13: Decimator / Interpolator

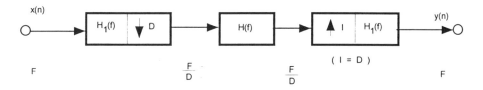

Fig. 1.14: As Fig. 1.13, with an intermediate filter

The link joining the decimator to the interpolator in Fig. 1.13 is operating at the slow sampling rate of F/D. This represents a convenient point for any further signal processing operations to be performed. Consider placing an intermediate filter joining the decimator to the interpolator as shown in Fig. 1.14. The reason for the intermediate filter is to allow freedom to choose D, which can now take any integer value but it must satisfy the system's filter specification. This can result in an even more computationally efficient structure than that in Fig. 1.13.

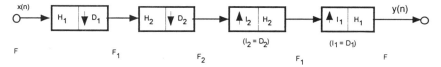

Fig. 1.15: Double decimation / double interpolation

The intermediate filter can also be implemented as a decimating/ interpolating filter, Fig. 1.15. Such a structure is called a multistage decimator/interpolator. The decimation is done in two stages, but the total decimation ratio must be identical with the one stage design, if the same filter specification is to be met. This idea can be extended to a larger number of stages. It has been shown in [33], that the computational rate is reduced with the number of stages. But the design and implementation becomes more complex and it is rarely justified to go for more than three stages.

APPENDIX 1.1: FREQUENCY SPECTRUM, FORM 1

The sampling process in Fig. 1.1(a) is modelled by the following multiplication

$$x_s(t) = x(t) \cdot p(t) = x(t) \cdot \sum_{n=-\infty}^{\infty} \delta(t - nT) \qquad (A1.1)$$

The spectrum of the sampled waveform, $x_s(t)$, is given by the Fourier transform (Appendix 1.3) as follows

$$X_s(\omega) = \int_{-\infty}^{\infty} x_s(t) e^{-j\omega t} \, dt$$

$$= \int_{-\infty}^{\infty} \left[x(t) \cdot \sum_{n=-\infty}^{\infty} \delta(t - nT) \right] e^{-j\omega t} \, dt \quad .$$

The order of integration and summation can be interchanged, hence

$$X_s(\omega) = \sum_{n=-\infty}^{\infty} \left[\int_{-\infty}^{\infty} \left(x(t) e^{-j\omega t} \right) \delta(t - nT) \, dt \right] \qquad (A1.2)$$

Applying the delta function sifting property (Appendix 1.3), we have

$$X_s(\omega) = \sum_{n=-\infty}^{\infty} x(nT) e^{-jn\omega T} \qquad , \qquad (A1.3)$$

where $x(nT)$ is a sample of $x(t)$ at the time $t = nT$.

APPENDIX 1.2: FREQUENCY SPECTRUM, FORM 2

Since the sampling sequence $p(t)$ is periodic, it can be expressed in terms of the Fourier series expansion (Appendix 1.3)

$$\left. \begin{array}{l} p(t) = \displaystyle\sum_{n=-\infty}^{\infty} c_n \, e^{-jn\omega_s t} \\[2em] c_n = \dfrac{1}{T} \displaystyle\int_{-T/2}^{T/2} p(t) e^{jn\omega_s t} \, dt \end{array} \right\} \qquad , \qquad (A1.4)$$

where T is the repetition period, and $\omega_s = 2\pi/T$ is the sampling frequency. Over the interval $(-T/2 \leq t \leq T/2)$, the sampling sequence $p(t)$ is represented by only the single Dirac impulse at $t = 0$, ie $\delta(t)$. Therefore, we have

$$c_n = \frac{1}{T} \int_{-T/2}^{T/2} \delta(t) e^{jn\omega_s t} \, dt = \frac{1}{T} e^{j0} = \frac{1}{T} \,,$$

where we have used again the sifting property (Appendix 1.3), with $t_0 = 0$. Hence, the alternative expression for $p(t)$ is given by

$$p(t) = \frac{1}{T} \sum_{n=-\infty}^{\infty} e^{-jn\omega_s t} \,.$$

We can now express the sampling operation as

$$x_s(t) = x(t) \cdot p(t)$$

$$= x(t) \cdot \left[\frac{1}{T} \sum_{n=-\infty}^{\infty} e^{-jn\omega_s t} \right]$$

or

$$x_s(t) = \frac{1}{T} \sum_{n=-\infty}^{\infty} x(t) \, e^{-jn\omega_s t}$$

Taking Fourier transform of the above, we have

$$X_s(\omega) = \frac{1}{T} \int_{-\infty}^{\infty} \left[\sum_{n=-\infty}^{\infty} x(t) \, e^{-jn\omega_s t} \right] e^{-j\omega t} \, dt \,,$$

which can be written as

$$X_s(\omega) = \frac{1}{T} \sum_{n=-\infty}^{\infty} \left[\int_{-\infty}^{\infty} x(t) e^{-j(\omega + n\omega_s)t} \, dt \right] \,,$$

where we have interchanged the order of summation and integration. The r-th term of the sum produces

$$X_{s,r}(\omega) = \int_{-\infty}^{\infty} x(t) \, e^{-j(\omega + r\omega_s)t} \, dt$$

$$= X(\omega + r\omega_s) \,,$$

which is the original analogue signal spectrum shifted from $\omega = 0$ to $\omega = -r\omega_s$. The total spectrum is the sum of these shifted versions and is given by

$$X_s(\omega) = \frac{1}{T} \sum_{r=-\infty}^{\infty} X(\omega + r\omega_s) \quad . \tag{A1.5}$$

APPENDIX 1.3: DIRAC FUNCTION; FOURIER PAIRS

The unit impulse function or the **Dirac delta function** is defined as

$$\delta(t - t_0) = \begin{cases} \infty, & t = t_0 \\ 0, & t \neq t_0 \end{cases}$$

and $\displaystyle\int_{-\infty}^{\infty} \delta(t - t_0)\, dt = 1 \quad .$

The delta function **sifting property**

$$I = \int_{-\infty}^{\infty} f(t)\, \delta(t - t_0)\, dt = f(t_0) \quad . \tag{A1.6}$$

Fourier transform pair for continuous-time aperiodic signals

$$\left. \begin{aligned} f(t) &= \frac{1}{2\pi} \int_{-\infty}^{\infty} F(\omega)\, e^{j\omega t}\, d\omega \\ F(\omega) &= \int_{-\infty}^{\infty} f(t)\, e^{-j\omega t}\, dt \end{aligned} \right\} \quad . \tag{A1.7}$$

Fourier series pair for continuous-time periodic signals

$$\left. \begin{aligned} f(t) &= \sum_{n=-\infty}^{\infty} c_n\, e^{jn\omega_0 t} \\ c_n &= \frac{\omega_0}{2\pi} \int_{-T/2}^{T/2} f(t)\, e^{-jn\omega_0 t}\, dt \end{aligned} \right\} \tag{A1.8}$$

where $\omega_0 = 2\pi/T$, and T is the repetition period.

APPENDIX 1.4: CONVOLUTION EQUATION DERIVATION

The convolution equation is usually derived in the time domain using the superposition principle. It is convenient here to derive it from the transfer function Eq. (1.25). We begin by setting $a_m = 0$, so that the output is given by

$$Y(z) = H(z) X(z)$$

$$= \left(\sum_{m=0}^{N} b_m z^{-m} \right) X(z) \quad .$$

For the unit sample input $x(n) = \delta(n)$, and $X(z) = 1$. This means that

$$Y(z) = H(z) = \sum_{m=0}^{N} b_m z^{-m} \qquad (A1.9)$$

and hence b_m's are the impulse response, i.e. $b_m = h(m)$.

For a general input signal $x(n)$, we have $X(z) = \sum_{i=0}^{\infty} x(i) z^{-i}$, and the

output is now

$$Y(z) = H(z) \cdot X(z)$$

or

$$\sum_{i=0}^{\infty} y(i) z^{-i} = \left[\sum_{m=0}^{N} h(m) z^{-m} \right] \left[\sum_{i=0}^{\infty} x(i) z^{-i} \right]$$

or

$$y(0) + y(1) z^{-1} + \cdot + y(n) z^{-n} + \ldots = \left[\sum_{m=0}^{N} h(m) z^{-m} \right]$$

$$\cdot \left[x(0) + x(1) z^{-1} + \ldots + x(k) z^{-k} + \ldots \right]$$

Taking the n-th term on the l.h.s. and for $m + k = n$ on the r.h.s. we have

$$y(n) z^{-n} = \sum_{m=0}^{N} h(m) \ x(k) z^{-(m+k=n)} \quad .$$

For $n = m + k$ we have $k = n - m$, which enables us to eliminate the variable k, so that

$$y(n) = \sum_{m=0}^{N} h(m) \ x(n - m) \qquad (A1.10)$$

,

which is the convolution equation.

Consider the transfer function (1.25) with $a_m \neq 0$. Now, the denominator polynomial roots (poles for H(z)) produce infinite length H(z). This can be seen, for example, by considering a single pole $p = - a_1$, which produces

$$\frac{1}{1 - a_1 z^{-1}} = 1 - a_1 z^{-1} + a_1^2 z^{-2} - a_1^3 z^{-3} + \ldots$$

Therefore, the numerator summation will extend to infinity, hence the convolution equation in this case will be

$$y(n) = \sum_{m=0}^{\infty} h(m) x(n - m) \qquad (A1.11)$$

2

The fundamental features of a DSP

2.0   INTRODUCTION

The basic concept of the digital signal processor has been encountered in Chapter 1 (see the block marked DSP in Fig. 1.1). The DSP needs to read the stream of input data samples ($x_b(n)$ in Fig. 1.1a), process them and send the output data stream to the DAC. The critical functions of the DSP can be seen by looking at the fundamental difference equation (1.24). The summation operations imply addition, while items being summed (accumulated) are products from multiplications involving stored input and output values. It will also be seen that the multiply-accumulate of stored items must be performed N times during the time delay between samples. At practical sampling rates of 10k to 100k samples/s, this requires a very fast processor with particularly efficient multiplication, addition and stored data movement. Many other important equations will be found in this book, e.g. the **convolution** of Eq. (1.29), which can be computed by the fast repetition of **multiply-accumulate** operations on stored data.

The traditional stored-program computer has a von Neumann architecture and a complex instruction set (CISC). In a von Neumann computer, the same memory is used to store both data and the program (machine code) which controls the operation of the machine. Consequently, the machine code for every instruction as well as each data value must be transferred sequentially on a single data bus. This creates a processing bottleneck which cannot be tolerated in a DSP. A CISC machine may take several clock cycles to execute one instruction (like a multiplication). Again, this is intolerable in a DSP. In contrast, in a reduced instruction set computer (RISC), instructions are few but every instruction is executed in a single clock cycle.

In this chapter, we shall look at the features of a practical but generalised digital signal processor. Section 2.1 describes the evolution of the DSP, Section 2.2 describes number representation and 2.3 shows how the crucial multiply-accumulate function can be implemented efficiently. The addressing of data in memory by 'normal' methods described in 2.4 and 2.5, gives addressing modes which are not found outside the DSP. We end with Section 2.6 describing the hardware-based schemes for fast iteration which are essential to all DSPs.

Many essential features of commercial DSPs can only be described in the context of a particular architecture and we shall describe some of these in Chapter 3, which can be regarded as an extension of Chapter 2.

2.1   DSP HISTORY AND PERFORMANCE

Ever since the first practical computers were developed in the 1940s, the general purpose CISC von Neumann computer has dominated the scene. This is still true today in spite of challenges from RISC designs, multiprocessors, etc. etc. Around 1984, when the first commercial DSPs were developed, a typical general purpose microprocessor had the following characteristics:

- Single memory for both program and data
- 8 bit data bus
- Complex instruction set (CISC)
- Fastest instruction executed in 1 microsecond
- 16 bit × 16 bit multiply in about 1 millisecond.

YEAR	RISC	ADD		MULTIPLY		ONE STEP OF CONVOLUTION	
		TIME (ns)	MACHINE CYCLES	TIME (ns)	MACHINE CYCLES	TIME (ns)	MACHINE CYCLES
1984	TMS32010	200	1	200	1	200	≈ 1
	TMS32020	200	1	200	1	200	≈ 1
	TMS320C25	100	1	100	1	100	≈ 1
1994	TMS320C50	30	1	30	1	30	≈ 1

(a)

YEAR	CISC	ADD		MULTIPLY		ONE STEP OF CONVOLUTION	
		TIME (ns)	MACHINE CYCLES	TIME (ns)	MACHINE CYCLES	TIME (ns)	MACHINE CYCLES
1984	8086	750	3	29500	118	43750	175
	80286	250	2	2625	21	4625	37
	80386	80	2	880	22	1120	28
1994	80486	17	1	433	26	493	29

(b)

Fig. 2.1:

(a) DSP performance   (b) CISC performance

For addition, multiplication and convolution

The multiply is particularly slow, because it is performed in software. Every instruction executed by a processor is carried out by a number of operations, each of which takes one machine cycle. In 1984, DSPs had a machine cycle time of about 200 ns and performed a multiplication in one or two machine cycles. Although general purpose CISC machines have become faster, they still cannot compete with DSPs on the sheer raw speed of simple arithmetic operations. Fig 2.1 compares add, multiply and convolution times for a typical integer DSP family (TMS320) and CISC family of processors over the years. Exact comparisons are not possible; for example CISC timings assume that values are stored in processor registers and would be considerably slower if values came from memory, as with the DSPs. However the figures give some idea of the relative performance. Note that the convolution calculations require some interation control, i.e. a loop counter must be tested at each multiply-accumulate step.

The term RISC requires some further explanation. A CISC (complex instruction set computer) processor executes a large number of different instructions each of which is thought to be useful for a particular programming task. A typical CISC processor instruction takes perhaps half a dozen machine cycles to execute. In contrast, a RISC (reduced instruction set computer), in theory executes a minimum set of instructions but each RISC instruction executes very quickly - usually in one machine cycle. Complex functions are computed by combining RISC instructions. This has led to the concept that machines supporting multi-cycle instructions can be described as CISC whereas those using mostly single cycle instructions, like DSPs, may be called RISC, **even if they support a large instruction set**.

Let us examine the efficiency of such a processor in executing an archetypal signal processing algorithm such as the convolution equation (1.29) from Chapter 1, reproduced here for N samples as Eq. (2.1).

$$y(n) = \sum_{m=0}^{N-1} h(m)x(n-m) \quad . \tag{2.1}$$

Similar types of multiply-accumulate operation, using two arrays occur in many signal processing situations, such as FIR filters and even PID (Proportional, Integral, Differential) control implementations. Speed of execution is of the essence, especially if the processor is to perform filtering tasks at the commercially important audio frequencies with, let us say, a 10 kHz sampling rate. The execution time is completely dominated by the multiplication in a conventional processor. Even if we could design a von Neumann-CISC processor with, say, a 100ns cycle time and a single cycle multiply, the von Neumann processor would typically have to carry out the following operations to compute each step of equation (2.1).

- fetch instruction, read operand 1, h(m) from memory
- fetch instruction, read operand 2, x(n-m) from memory
- fetch instruction, multiply operands
- fetch instruction, accumulate product
- fetch instruction, decrement & test counter, jump

These operations have to be carried out N times, which requires iteration control, and every instruction must be fetched **from the same memory that holds the data**. Thus it is not likely that the five operations shown above could be executed in less than about 10 machine cycles by a von Neumann processor (assuming operands are read from memory and multiply and accumulate operate on registers). If integer representation is used, most signal processing arithmetic requires a word length of at least 16 bits. In order to avoid the time penalty of multiple memory accesses therefore, the processor word length must be at least this value. It is currently possible to execute such a multiply-accumulate function at a rate of almost one machine cycle per multiply-accumulate by using non-von Neumann processor designs. A machine cycle can currently be performed in under 30 ns at the time of writing and easily allows the implementation of audio frequency digital filters, for example.

2.2   NUMBER REPRESENTATION IN DSPS

The signal processing described in Chapter 1 (see Fig. 1.1) implies that each data sample is represented as a binary word whose bits are processed concurrently. All the digital signal processors described in this book do this, in fact. It should be noted, however that bit-serial arithmetic, where bits are processed serially used to be popular and is again becoming used in gate array-based ASIC devices.

2.2.1   Integer number representation

Following normal computing practice, binary number representation is almost always used in digital signal processing arithmetic. Such numbers may be considered as **natural binary** integers with N bits representing $2^0, 2^1 \ldots \ldots \ldots 2^{N-1}$. Therefore an N-bit register holds integer values from 0 to $(2^N - 1)$. This is suitable for many situations, such as loop counters or address pointers for instance where all values are positive integers. In a digital sample however, it is necessary to represent both positive and negative values. The most convenient and efficient approach to handling negative values is to use twos complement representation where an N-bit register uses the natural binary range of 0 to $(2^{N-1} - 1)$ to represent positive values naturally and the range 2^{N-1} to $(2^N - 1)$ to represent negative values. A negative value m is represented in N-bit twos complement as $2^N + m$, or $2^N - M$, where M is the **magnitude** of m, |m|, i.e. M > 0. The result of this is that natural

Each digital sample of the analogue signal must describe its polarity and magnitude. COMPUTATION	TWOS COMPLEMENT REPRESENTATION	N-BIT REGISTER IS $\geq 2^N$ (BIT N LOST)
P1 + P2	P1 + P2	
P1 + m2	P1 + $(2^N - M2)$	P1 - M2 $[\geq 0]$
m1 + m2	$(2^N - M1) + (2^N - M2)$	$2^N - (M1 + M2)$ $[< 0]$
P1 - P2	P1 - P2	
P1 - m2	P1 - $(2^N - M2)$	
m1 - m2	$(2^N - M1) - (2^N - M2)$	
m1 - P2	$(2^N - M1) - P2$	

Each digital sample of the analogue signal must describe its polarity and magnitude. COMPUTATION	N-BIT REGISTER IF $0 \leq$ REGISTER $< 2^N$	N-BIT REGISTER IF REGISTER < 0 (BORROW 2^N)
P1 + P2	P1 + P2 $[\geq 0]$	
P1 + m2	$2^N - (M2 - P1)$ $[< 0]$	
m1 + m2		
P1 - P2	P1 - P2 $[\geq 0]$	$2^N - (P2 - P1)$ $[< 0]$
P1 - m2		P1 + M2 $[\geq 0]$
m1 - m2	M2 - M1 $[\geq 0]$	$2^N - (M1 - M2)$ $[< 0]$
m1 - P2	$2^N - (M1 + P2)$ $[< 0]$	

Fig. 2.2: Natural binary addition and subtraction of twos complement positive values (P) and negative values (m) in an N-bit register: $0 \leq P \leq (2^{N-1}-1)$; $-2^{N-1} \leq m \leq -1$; M = |m|. The sign of actual results is given in square brackets.

COMPUTATION	TWOS COMPLEMENT REPRESENT-ATION	N-BIT REGISTER IS $\geq 2^N$ (BIT N LOST)	N-BIT REGISTER IF $0 \leq$ REG $< 2^N$	N-BIT REGISTER IF REG. < 0 (BORROW 2^N)
2 + 3	0010 + 0011		0101 [5]	
2 + (-1)	0010 + 1111	0001 [1]		
2 + (-3)	0010 + 1101		1111 [-1]	
(-3) + (-2)	1101 + 1110	1011 [-5]		
3 - 2	0011 - 0010		0001 [1]	
2 - 3	0010 - 0011			1111 [-1]
4 - (-2)	0100 - 1110			0110 [6]
-1 - (-3)	1111 - 1101		0010 [2]	
-3 - (-1)	1101 - 1111			1110 [-2]
-3 - 2	1101 - 0010	1011 [-5]		

Fig. 2.3: Examples of twos complement arithmetic in a 4 bit register (N = 4, compare with Fig. 2.2) Values in square brackets are decimal equivalents

binary operations, such as add and subtract can be used directly on twos complement values, as shown in Fig. 2.2. The values are represented in twos complement and the computation carried out in natural binary in an N-bit register. If the computed result is $\geq 2^N$, the most significant bit is lost from the N-bit register. If it is < 0, then 2^N has to be 'borrowed', i.e. added, in order to carry out the subtraction. As shown in Fig. 2.2, this gives the correct result R where a positive value of R is in natural binary and a negative one is given by $2^N - |R|$. Fig 2.3 gives examples in a 4-bit register ($2^N = 16$ decimal $= 10000$ binary) of all the cases described in Fig. 2.2. Note that the result will only be correct if it is within the range set by the register size. Overflow (or underflow) will occur unless $(-2^{N-1}) \leq R < 2^{N-1}$. Twos complement overflow has particularly catastrophic results in a DSP environment because if the register contents cross the boundary from $(2^{N-1}-1)$ to 2^{N-1} the twos complement equivalent changes from a large positive to a large negative number.

2.2.2   Sign extension

All twos complement values $\geq$ zero have a most significant bit (MSB) of 0 and all values $<$ zero have a MSB of 1 because the MSB represents 2^{N-1}. Thus the MSB is known as the sign bit. A 16-bit register is the minimum that is useful in DSP work and can represent -32768 to +32767; a 24 bit register may represent -8388608 to +8388607. These correspond to 90 dB and 138 dB dynamic range respectively.

Integer DSP devices normally perform arithmetic in an accumulator register of at least twice the normal word size, e.g. the Motorola DSP56000 DSP represents numbers in a 24 bit word but performs arithmetic in a 56 bit accumulator. In order to maintain the correct sign therefore, a twos complement value needs to be sign extended if stored in a long register, i.e. the upper part of the long register must contain a copy of the sign bit, rather than be filled with zeros. The sign bit may, of course, be zero but may be one. All integer DSPs can perform this **sign extension** in hardware without any time penalty. Fig 2.4 gives some examples of sign extended and zero extended values.

16 BIT HEX VALUES	ZERO EXTENDED TO 32 BITS GIVES INCORRECT NEGATIVE VALUES		SIGN EXTENDED TO 32 BITS ALL VALUES ARE CORRECT	
7FFF [+32767]	00007FFF [+32767]	√	00007FFF [+32767]	√
8001 [-32767]	00008001 [+32769]	×	FFFF8001 [-32767]	√
0001 [+1]	00000001 [+1]	√	00000001 [+1]	√
FFFF [-1]	0000FFFF [+65535]	×	FFFFFFFF [-1]	√

Fig. 2.4: Twos complement must be sign extended when moved to a longer register. Decimal equivalents are in square brackets

2.2.3   Fractional representation

In straightforward natural binary or twos complement binary, there is no binary point, in other words, the binary point may be considered to be to the right of the least significant bit, which represents 2^0. When performing arithmetic on twos complement integer values, it is usually conceptually more convenient to divide an N-bit binary value by 2^{N-1} to represent a value less than 1; this is known as

	BINARY	FRACTION	DECIMAL
(a)	01010000 + 00100000 = 01110000	80/128 + 32/128 =112/128	0.625 + 0.250 = 0.875

(b)	10110000 + 01100000 = 00010000	- 80/128 + 96/128 = 16/128	- 0.625 + 0.750 = 0.125

(b)

Fig. 2.5:
(a) 0.625 + 0.25 = 0.875
(b) -0.625 + 0.75 = 0.125
Twos complement addition of fractions carried out in integer arithmetic

fractional representation. Thus, in this fractional representation, the binary point is considered to be immediately to the right of the sign bit. An 8-bit binary value would therefore represent values between -128/128 and +127/128 and a 16-bit value -32768/32768 to +32767/32768.   Fig 2.5 demonstrates the addition of two fractions, using 8 bit values.

   The advantage of treating binary numbers as fractions between +1 (almost) and -1 is most clearly seen in multiplication. Multiplying two **unsigned** N bit binary numbers gives a 2N bit product. The multiplication of two N bit **twos complement** numbers gives a 2N-1 bit product, as shown in Fig. 2.6. In a practical processor, this is normally stored in a 2N bit register. If fractional representation is used, this can *always* be scaled to give a correct N-bit answer by left-shifting by one bit and taking the most significant N bits.

BINARY	FRACTION	DECIMAL
01100000	96/128	0.750
x 01000000	x 64/128	x 0.500
= 00110000	= 48/128	= 0.375

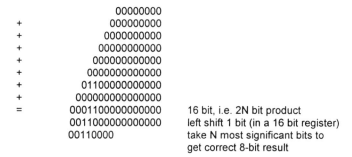

```
           00000000
 +        000000000
 +       0000000000
 +      00000000000
 +     000000000000
 +    0000000000000
 +    01100000000000
 +    000000000000000
 =    0001100000000000     16 bit, i.e. 2N bit product
      0011000000000000     left shift 1 bit (in a 16 bit register)
      00110000             take N most significant bits to
                           get correct 8-bit result
```

Fig. 2.6: Long multiplication of two 8 bit fractions

The above scheme of treating the input operands to the multiplication as fractions is not a mandatory system. The benefits are conceptual, rather than physical. However, if twos complement N bit operands are treated as integers and not fractions, taking the least significant N bits of the product (the natural thing to do) may lose information in the more significant bits. Most DSPs have the capability to produce such a left shift, usually without any time penalty. Note that a twos complement multiplier behaves differently from an unsigned multiplier. Unsigned multiplication may sometimes be necessary, e.g. for double precision arithmetic and integer DSPs normally support both, these days.

2.2.4   Floating point representation

The problems of limited dynamic range and the need for the user to keep track of scaling manually can be eliminated by using a floating point representation of the samples. The IEEE754 standard [14] describes a 32 bit floating point standard which represents each value by a 23 bit twos complement mantissa, a sign bit and an 8-bit exponent as shown in Fig. 2.7.

The sign bit and the mantissa represent a 25 bit twos complement value between -2 and $+2-2^{-24}$, in other words the binary point is preceded by a sign bit and one binary digit. The mantissa is preferably normalised, i.e. has a value greater than 1 or less than -1. This means that all values start with either 01.......... (positive) or 11......... (negative) Because the bit preceding the binary point is always 1, it is not stored. There is a whole group of DSPs which can handle 32 bit floating point numbers in IEEE754 or similar formats. This effectively eliminates the problems of scaling with the penalty of considerable system cost. A floating point DSP inevitably needs a 32 bit data bus (at least) implying higher memory costs and integer DSPs are to be preferred in high volume applications where the dynamic range permits.

SEEEEEEEEMMMMMMMMMMMMMMMMMMMMMMMM

S = sign bit
E = exponent bit
M = mantissa bit

Fig. 2.7: 32 bit IEEE 754 floating point format

It is sometimes worthwhile to implement floating point in software on an integer DSP, especially as the instruction set of most integer DSPs supports operations, such as efficient shifting (for normalisation). Suitable formats are processor dependent and more details and a suite of floating point routines for the TMS320C25/C50 can be found in Appendix 2.1.

2.3   THE MULTIPLY-ACCUMULATE OPERATION

It was stated at the start of this chapter that DSP devices can perform multiply-accumulate operations, e.g. to compute Eq. (2.1), very efficiently. Unfortunately, although this is true, practical processors differ in how they do this and it is difficult to discuss fully this important DSP feature independently of a specific architecture. In fact, as we shall see in Chapter 3, processor architectures have to a large extent been dictated by the designers' approach to calculating a sum of products efficiently. There are however some essential principles which are common to all DSPs. A straightforward calculation of Eq. (2.1), using arrays to hold N coefficients and N samples of x, is described in a C-like pseudo-code in Fig. 2.8a. Input samples are stored in x[0] (newest) to x[N-1] (oldest) and coefficient values in h[0] to h[N-1]. Note that, with each iteration, the values of x are shuffled along the array so that the oldest sample is lost and the next sample can be entered.

`// arrays h[], x[] hold h and x data`
`Yn=0;`
`for (m=N-1;m>=0;m=m-1;){`
`    x[m+1] = x[m]; //move x data`
`    Yn = Yn + (h[m] * x[m]);`
`}`
`x[0] = next_sample;`
	`ph = &h[N-1];   px = &x[N-1]; pq = px+1;`
	`Yn=0;`
	`for (i=0;i<N;i=i+1;){`
	`    *pq-- = *px; // move x data`
	`    Yn = Yn + (*ph-- * *px--);`
	`}`
	`*px = next_sample;`

(a)	(b)

Fig. 2.8: Pseudo-code to compute the equation $y(n) = \sum_{m=0}^{N-1} h(m)x(n-m)$

(a) Using arrays
(b) Using address pointers

Fig 2.8b reorganises the code to use address pointers which makes use of a fast post-decrement addressing mode, implemented on all DSPs. The pointers can 'walk along' the array as the calculation progresses, thus avoiding the calculation of the address of each array element independently. A third pointer pq is used to copy the x values. Several processors (TMS320C25/50 & DSP32) use a similar algorithm to implement a convolution. All DSPs also use two data buses to transfer the h and the x values simultaneously to the multiplier-accumulator and can execute the algorithm described in Fig. 2.8b in almost N machine cycles. It is possible to avoid the physical movement of the x values by using circular addressing (see Section 2.5.2). As we shall see in Chapter 3, h and x are sometimes held in two clearly separate address spaces, each with its own data and address bus, so that these machines do not have the single von Neumann address space familiar in most general purpose machines. Some of the powerful floating point DSPs lay claim to a single von Neumann address space (Fig. 3.1). This is not entirely justified as two data buses are still necessary to read the two multiplicands in one machine cycle. Apart from fast pointer arithmetic, the remaining feature necessary for the fast execution of the multiply-accumulate is fast iteration. In the remainder of this chapter, we shall explore the address pointer and iteration schemes common to most DSPs. In the next chapter we shall look at implementation on the different architectures which are currently available.

2.4   FUNDAMENTAL DSP ADDRESSING MODES

DSPs do not possess the enormous variety of addressing modes used on modern general purpose processors. The principal addressing modes that will be familiar to users of conventional assembly languages are:

- Register addressing (register to register transfer)
- Direct addressing (memory location specified in instruction)
- Immediate addressing (constant included in instruction)
- Indirect addressing (register holds address)
- Post-increment/decrement by 1 (indirect + register modify)
- Post-increment/decrement by a value (indirect + register modify)

Speed is critical in digital signal processors and most of these addressing modes are executed in a single machine cycle on a DSP. Register addressing needs no further comment. Immediate addressing on most 'normal' processors takes two machine cycles, the first to fetch the instruction, the second to read the constant which is held in a second memory location. Similarly, direct addressing normally requires a whole word to hold the address where the data is located, in addition to the instruction word, again requiring two machine cycles to execute the instruction. However DSPs often use strange architectures or impose restrictions to increase the speed of immediate and direct addressing modes. The method used depends on the processor. The more interesting ones are described in Chapter 3 under specific processors. The inexperienced may be surprised to discover that instructions normally considered to be essential are often missing from DSPs. For example, the Texas Instruments TMS320C30 floating point processor has no immediate addressing mode to handle a full size (32 bit) value. Such values must be handled via data lists stored in memory at assembly-compile time.

The various indirect addressing modes form a pivotal part of the instruction set of any DSP and features common to virtually all designs will now be described.

2.4.1   Normal indirect addressing modes

In indirect addressing, the location in memory where the value is stored is held in an address pointer register. The essential reason for this is so that arithmetic can be performed on the address. The most common arithmetic operations are to increment or decrement the address pointer register by a value, often unity. A post-increment addressing mode will use the value stored at an address and afterwards modify the address pointer, just like the *<pointer>++ construct in the C language and without any speed penalty. Note: in the rest of this chapter 'increment' can be taken to include the decrement function. DSPs normally allow post-increment, not only by unity, but also by the value held in another register. This may be useful for stepping through a two-dimensional array, e.g. in a set of complex numbers held in the order

real1, imaginary1, real2, imaginary2, real3, imaginary3........

incrementing the address pointer by 2 enables the programmer to step through the list of real values at a rate of one per machine cycle. Unlike other processors, DSPs invariably allow post-increment address arithmetic to be performed in a single machine cycle concurrently with instruction execution. The computation of the basic convolution equation Eq. (2.1) and similar operations requires two arrays of values (h and x). It may therefore come as no surprise to find that DSPs incorporate address generation hardware which can update two address pointers simultaneously. Unfortunately a wide variety of methods is used. More detail about the capability of specific processors is given in Chapter 3.

2.5   SPECIAL DSP INDIRECT ADDRESSING MODES

There are some indirect addressing modes which are problem-specific and are found on virtually all modern DSPs but are never found in general purpose processors. We shall describe the principles here. Implementation details for a particular processor vary and some examples are given in the next chapter.

2.5.1   Bit-reverse (reverse carry) addressing

It will be shown in Section 5.2 that the coefficients of a FFT computed 'in place' normally appear in 'scrambled' rather than the natural order. The natural order may be restored by reversing the bits of the index of the array of coefficients. Almost all modern DSPs have a bit-reversed post-incrementing addressing mode which allows such coefficients to be accessed in their proper order without any time penalty. In bit reversed (or reverse carry) addressing, the addition of an address pointer to a value in post-increment addressing is carried out with the carry bit propagating in the reverse direction (towards bit 0). Let us take the 8 bit FFT illustrated in Fig. 5.4 as an example. We post-increment an address pointer by 4 (half the FFT length) but use reverse-carry arithmetic. Fig. 2.9 shows a buffer holding values in bit-reversed order and the result of repeated post-increments. The values are accessed in the natural order.

FFT COEFFICIENTS IN MEMORY		ADDRESS POINTER DURING BIT-REVERSE POST-INCREMENT ADDRESSING	
MEMORY ADDRESS	MEMORY CONTENTS	POINTER BITS	ADDRESS ACCESSED
A	x(0)	XXXX000	A+0
A+1	x(4)	XXXX100	A+4
A+2	x(2)	XXXX010	A+2
A+3	x(6)	XXXX110	A+6
A+4	x(1)	XXXX001	A+1
A+5	x(5)	XXXX101	A+5
A+6	x(3)	XXXX011	A+3
A+7	x(7)	XXXX111	A+7

Fig. 2.9: Reverse-carry indirect addressing used to access FFT coefficients

2.5.2   Circular addressing

The circular buffer is a concept which is well known to high level language programmers. For example, an area of memory is allocated as a buffer. Typically, two address pointers are maintained, one for writing data to the buffer, the other for reading buffer contents. After both read and write operations, the appropriate buffer pointer is incremented. If the incremented pointer exceeds the top buffer address, it is loaded with the lowest buffer address. Such a scheme is commonly used to implement a first-in-first-out buffer as in Fig. 2.10. General purpose computers invariably implement such buffers in software and incrementing, testing and loading the address pointers to achieve circulation from one end of the buffer to the other is a substantial overhead. Circular addressing can be used to avoid the movement of stored data and the repeated initialisation of the x pointer in the 'data movement' multiply-accumulate algorithm of Fig. 2.8b. The pointer pq and the x data move can be eliminated. Fig 2.11 compares the circular implementation with the 'data movement' version. After reading an input sample (and processing N values from the buffer), the address pointer ends up pointing to the location of the oldest x value. The latest x value replaces the oldest and the location of the latest sample is rotated round the circular buffer one position each

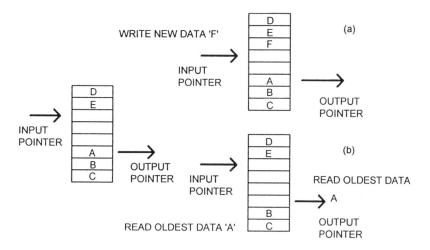

Fig. 2.10: First-in-first-out circular buffer storing data A (oldest) to E (newest)
(a) Write new data F
(b) Read oldest data A

ph = &h[N-1];   px = &x[N-1]; pq = px+1;
Yn=0;
for (i=0;i<N;i=i+1;){
    *pq-- = *px; //move x data
    Yn = Yn + (*ph-- * *px--);
}
*px = next_sample;

ph = &h[N-1];
Yn=0;
for (i=0;i<N;i=i+1;){
    Yn = Yn + (*ph-- * *px--);
    if(px < &x[0]) px = &x[N-1]; //circulate x ptr
}
*px = next_sample;

    (a)   (b)

Fig. 2.11: Pseudo-code to compute the equation $y(n) = \sum_{m=0}^{N-1} h(m)x(n-m)$

(a) Using post-decrement addressing
(b) Using circular addressing

time.   Thus, compared to the general form of the FIFO buffer of Fig. 2.10, a single pointer can serve as both input and output pointer. If pointer circulation is implemented in software there is little speed advantage. However many DSPs implement some form of circular addressing in hardware with no speed penalty and the most modern DSPs tend to use this method to compute equation (2.1).

Some DSPs incorporate an addressing mode known as modulo addressing which allows circular buffers to be supported without a software overhead. In modulo addressing, only the least significant N bits of the address pointer are involved in address arithmetic. The upper bits of the pointer remain fixed. Thus the pointer can only address locations between a base address B and address B + 2^N - 1. Thus post-incrementing such a pointer by one can be used to access all the locations in a circular buffer sequentially. It is possible to use a post- increment value greater than 1 in modulo addressing to allow two-dimensional arrays to be accessed in row order or column order without any time penalty for pointer manipulation. Fig 2.12 gives an example for a buffer of size 8, incremented by 2.

ADDRESS POINTER VALUE	POINTER BITS 0-2 DURING A MODULO 8 INCREMENT BY 2 UPPER BITS REMAIN FIXED
A	xxxx000
A+2	xxxx010
A+4	xxxx100
A+6	xxxx110
A	xxxx000 etc.

Fig. 2.12: Modulo 8 addressing

Increment by two could be useful with interleaved real and complex numbers. The disadvantage of modulo addressing is that the programmer does not have complete freedom to locate the buffer anywhere in memory. It must start on a page boundary, so that it may be necessary to allocate an area of memory of twice the buffer size, making large buffers impractical.

There is no fundamental reason why hardware support for a circular buffer should be limited to the modulo arithmetic scheme. Unrestricted circular addressing is available on the most recent DSP designs that operate like the scheme of Fig. 2.11b but where the pointer circulation is performed by hardware. The programmer stores the buffer start and end addresses and an out-of-range check is performed on the pointer in hardware.

2.6   SPECIAL DSP ITERATION SCHEMES

Because the core operation of many DSP algorithms is short, the iteration overhead, i.e. the code needed to control the repetition, is often a substantial proportion of the execution time. Note the *for* loop in Fig. 2.11; the complete iteration of Fig. 2.11 is itself repeated for each sample. Hardware support for iteration control is a part of all DSPs and can be categorised into

- The repetition of one instruction
- Repetition of a block of code.

2.6.1   Single instruction repeat

The repetition of a single instruction can be especially powerful in a DSP. If the instruction includes post-increment addressing, the repetition can step through one or more blocks of memory holding arrays of data. For example, the AT&T DSP32 processor uses the following **single instruction** to perform the basic multiply-accumulate described in Fig 2.8b:

$$a1 = a1 + (*r3++ = *r4++) \quad * \quad *r5++$$

This looks like and in fact is legal C, but is also a legal line of assembler for the AT&T processor (C was developed by AT&T) which translates to a single machine code instruction, where a1, r3, r4 and r5 are processor registers. The two instructions repeated in the *for* loop of Fig. 2.8b could have been written as one line of C:

$$Yn = Yn + (*pq-- = *px--) \quad * \quad *ph-- \qquad .$$

The similarity to the DSP32 instruction is obvious and is just one example of the power of the efficient multiple pointer arithmetic found in all DSPs.

When repeating a single instruction like this, it is possible to avoid the instruction fetch operation for N-1 of the N repetitions. DSPs usually incorporate a REPEAT N type of instruction which repeats the instruction that follows N times. The execution of a REPEAT instruction will load a hardware counter with N. The following instruction is fetched once and executed N times. The decrement and test of the counter is also performed in hardware at each iteration. It is this capability that allows the DSP to multiply-accumulate at the rate of one sample per machine cycle and is used in the FIR filter implementations in this book.

One of the disadvantages of this type of single instruction repeat is that the complete iteration is treated as one instruction and interrupts cannot normally be accepted until the finish. This may cause the worst case response to an interrupt to be unacceptably long. A solution may be to split up a long iteration into two or more hardware supported REPEAT loops.

2.6.2   Multiple instruction repeat

More recently, DSP designers have incorporated zero overhead iteration, based on hardware, for a block of instructions. Detailed implementations vary but Fig. 2.13 describes the principles. At the start of the repeated block, an instruction loads three iteration control registers with

(1) The address of the next instruction (start of loop)
(2) The address of the end of the loop
(3) Number of iterations

The block to be repeated then executes. As each instruction is executed, the program counter is compared with the address of the end of the loop. Each time that the end of the loop is encountered, the loop counter is decremented and the program counter loaded with either the start of the loop or the following instruction as described in Fig. 2.13.

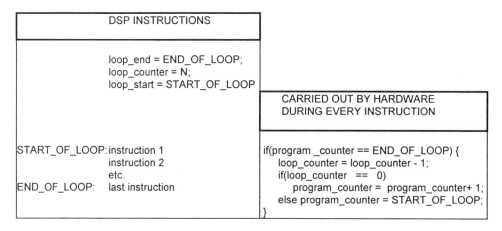

Fig. 2.13: DSP Hardware to repeat a block of instructions

Although the instruction fetches cannot be avoided in this case, the counter decrement and test can be performed in hardware while an instruction is being executed. No instructions to decrement the counter, test its value and jump to the start of the loop are necessary.

Most processors allow interrupts during the execution of a repeated block of instructions. If iterative loops are nested, obviously the greatest speed gain is achieved by using the hardware supported loop control for the innermost one. However it is practicable to save the iteration control registers, usually on a stack to implement nested hardware controlled loops. Some DSPs perform this operation in hardware also to allow even faster nested loops.

2.6.3   The delayed branch

	NOT FULL SPEED			FULL SPEED ATTAINED				
FETCH	A	B	C	D	E	F	G	H
DECODE		A	B	C	D	E	F	G
READ			A	B	C	D	E	F
EXECUTE				A	B	C	D	E

Fig. 2.14: Pipelined execution of 8 instructions A to H:
full speed attained only when fetch, decode, read & execute are carried out concurrently

MEMORY LOCATION	INSTRUCTION	EXECUTED?
1	A	YES
2	B (NORMAL BRANCH TO INSTR H)	YES
3	C	
4	D	
5	E	
6	F	
7	G	
8	H	YES
9	I	YES
10	J	YES

(a)

	FULL SPEED NOT ATTAINED								FULL
FETCH	A	B	C	D	E	H	I	J	K
DECODE		A	B	-	-		H	I	J
READ			A	B	-			H	I
EXECUTE				A	B				H
CYCLE	1	2	3	4	5	6	7	8	

(b)

Fig. 2.15: Pipelined normal branch instruction execution showing pipeline break
(a) List of instructions
(b) Steps during execution

Any iteration involves a branch instruction to load the program counter with an address which is out of sequence. Most DSPs use a pipelined system for fetching and executing instructions. This means that every instruction is executed in a sequence of steps, e.g. fetch, decode, read data, execute. If maximum speed is to be attained, all four steps of four different instructions must be executed concurrently, as shown in Fig. 2.14. If an instruction to branch to another address is executed, this breaks the pipeline and prevents the attainment of maximum speed. An example of a normal branch is shown in Fig. 2.15. In this case instructions C, D and E would be fetched during the execution of the branch instruction B even though they are not going to be executed. This can be avoided with a delayed branch instruction as shown in Fig. 2.16. In this case, the three instructions C, D, E after the delayed branch instruction dB are executed and the pipeline is kept full. Obviously the delayed branch is rather unfriendly to the programmer who must remember that the branch instruction does not mark the 'jump' to a new address. It is particularly unfriendly to compiler writers who may not be able to ascertain if a delayed branch can be used.

MEMORY LOCATION	INSTRUCTION	EXECUTED ?
1	A	YES
2	dB (DELAYED BRANCH TO INSTR H)	YES
3	C	YES
4	D	YES
5	E	YES
6	F	YES
7	G	
8	H	YES
9	I	YES
10	J	YES

(a)

			FULL SPEED ATTAINED							
FETCH	A	dB	C	D	E	F	G	H	I	J
DECODE		1	dB	C	D	E	F	G	H	I
READ			1	dB	C	D	E	F	G	H
EXECUTE				1	dB	C	D	E	F	H
CYCLE	1	2	3	4	5	6	7	8		

(b)

Fig. 2.16: Delayed branch instruction execution showing unbroken pipeline
(a) List of instructions
(b) Execution of instructions

2.7   SUMMARY

In this chapter we have examined the distinctive features of DSPs which distinguish them from other types of processor. Not all these features are found in all DSPs but you can be sure any new devices are likely to use most of them. Chapter 3 will compare the architectures of some commercial DSPs and summarise the features of currently available devices that have been mentioned above. Many interesting architectural features are so specialised that they can only be described in terms of a particular manufacturer's device; these are also outlined in chapter 3.

APPENDIX 2.1:

TMS320C25/C50 floating point add, subtract, multiply & divide


```
;TMS320C25/c50 floating point + - * / routines
;2 word floating point representation equivalent
;to a 31 bit integer
;high word = 16 bit mantissa + or - 1
;low word = 4 bit exponent 0 to 15
;e.g. 0.5, mantissa=16384, exponent=15
                include 'h32025.inc'
;
                ramaddr 300h
v1              ramspace 2              ;operand 1
v2              ramspace 2              ;operand 2
result          ramspace 2              ;place result here
temp            ramspace 1              ;temporary store

                ldpk h.v1
cont            call fpdiv
stop            b cont

fpadd                                   ;result = v1 + v2 (10-26 cycles)
                lt l.v1                 ;v1 exponent
                lact l.v1 + 1           ;v1 mantissa
                lt l.v2                 ;v2 exponent
                addt l.v2 + 1
                lark ar1,16             ;count down exponent from 16
                larp ar1
                rptk 15                 ;shift 1 to 16x
                norm *-                 ;unless overflow
                sach l.result + 1,noshift
                sar ar1,l.result
                ret
;
fpsub           ;result = v1 - v2
                lt l.v1                 ;v1 exponent
                lact l.v1 + 1           ;v1 mantissa
                lt l.v2                 ;v2 exponent
                subt l.v2 + 1
                lark ar1,16             ;count down exponent from 16
                larp ar1
                rptk 15                 ;shift 1 to 16x
                norm *-                 ;unless overflow
                sach l.result + 1,noshift
                sar ar1,l.result
                ret
```

```
;
fpmul           ;result = v1 * v2 (13-28 cycles)
                lar ar0,l.v1            ;v1 exponent
                lar ar1,l.v2            ;v2 exponent
                larp ar1
                mar *0+                 ;ar1= exp1 + exp2
                sbrk 15                 ;due to multiply shift
                lt l.v1 + 1
                mpy l.v2 + 1            ;P = mantissa product
                pac
                rptk 15                 ;shift 1 to 16x
                norm *-
fpmx            sach l.result + 1,noshift
                sar ar1,l.result
                ret
;
fpdiv                                   ;result = v1 / v2 (24-52 cycles)
                lar ar0,l.v2            ;v2 exponent
                lar ar1,l.v1            ;v1 exponent
                larp ar1
                mar *0-                 ;ar1= exp1 - exp2
                adrk 16
                lac l.v2 + 1,noshift    ;v2 mantissa
                abs                     ;must be +ve for subc
                sacl l.temp,noshift
                lac l.v1 + 1,shift15    ;v1 mantissa/2
                abs                     ;must be +ve for subc
                rptk 14                 ;divide 15 bits
                subc l.temp             ;divide
                sacl l.temp,noshift
                zalh l.temp
                rptk 14                 ;shift 0 to 15x
                norm *-
                sach l.result + 1,noshift
                sar ar1,l.result
x               lt l.v1  + 1   ;adjust sign
                mpy l.v2 + 1
                pac
                bgez fpdx
                zalh l.result + 1
                neg
                sach l.result + 1,noshift
fpdx            ret
```

3

The Practical DSP devices

3.0   INTRODUCTION

This chapter is concerned with practical commercial processors and may be regarded as an annex to Chapter 2. We shall briefly review the architectures of four commercial integer and four floating point digital signal processors. The intention is not to give a complete description for which the appropriate manufacturer's user guide is the best source. Rather, we will point out that most DSPs use some unusual and interesting features to improve their performance.

processor	DSP56000	TMS320C25	TMS320C50	ADSP2100
manufacturer	Motorola	Texas Instruments	Texas Instruments	Analog Devices
bits/word	24 + 24 + 24	16 + 16	16 + 16	24 + 16
memory size (words)	4M + 4M + 4M	64K + 64K	64K + 64K	16K + 16K
memory spaces	2xdata+prog	prog + data	prog + data	prog + data
on-chip memory	256X + 256Y	512 (data)	1K (data)	0 + 0
on-chip stack size	16 words	8 words	8 words	16 words
external buses	1	1	1	2
machine cycle	80ns	100 ns	35 ns	80 ns

(a)

processor	DSP96000	TMS320C30	DSP32C	ADSP21020
manufacturer	Motorola	Texas Instruments	AT&T	Analog Devices
bits/word	32 + 32 + 32	32	32	48 + 40
memory size (words)	4G + 4G + 4G	16M	16M	16M+4G
memory spaces	2xdata+prog	von Neumann	von Neumann	prog + data
on-chip memory	512X + 512Y	2K	1536	0 + 0
on-chip stack size	16 words	N/A	N/A	20 words
external buses	2	1	1	2
	60 ns	60 ns	40 ns	60 ns

(b)

Fig. 3.1: Basic architectures of some popular DSPs

(a) Integer

(b) Floating point

Many of these can only be described in the context of a specific processor and often illustrate an important principle. It is inevitable that any discussion of features of commercial products is likely to become out of date. DSPs are not changing as fast as they were a few years ago but the user might be on the lookout for faster cycle times as well as more on-chip memory and peripheral devices such as ADCs. From the practical DSPs in this section, the reader should be able to assess both current and future processor features.

We shall start in Section 3.1 by describing four categories in which these eight DSPs can conveniently be placed and making a few general observations. Interesting features of the basic architectures will be outlined in Sections 3.2 to 3.5. The multiply-accumulate operation has often been the basic drive behind these architectures and this is given prominent treatment, as it was in Chapter 2.

3.1   OVERVIEW OF COMMERCIAL DSPS

Fig. 3.1 gives an architectural summary of the DSPs used in this chapter. We have grouped the processors which only process integers separately from the floating point machines. The integer group of processors in 3.1(a) would usually be considered as commercial rivals, as would the floating point processors. The integer processors would be considered for cost-sensitive applications requiring little memory and the floating point DSPs for high performance, memory hungry tasks. However, it is convenient to group the processors according to architecture as four groups of two:

processor	DSP56000	TMS320C25	TMS320C50	ADSP2100
data word size	24	16	16	16
repeat 1 instruction	YES	YES	YES	YES
repeat block of instructions	YES	NO	YES	YES
addressing post-inc/dec by 1	YES	YES	YES	YES
addressing post-inc/dec by n	YES	YES	YES	YES
reverse carry addressing	YES	YES	YES	YES
modulo addressing	YES	NO	YES	NO
true circular addressing	NO	NO	YES	YES
delayed branch	NO	NO	YES	YES

(a)

processor	DSP96000	TMS320C30	DSP32C	ADSP21020
data word size	32	32	32	32
repeat 1 instruction	YES	YES	YES	YES
repeat block of instructions	YES	YES	YES	YES
addressing post-inc/dec by 1	YES	YES	YES	YES
addressing post-inc/dec by n	YES	YES	YES	YES
reverse carry addressing	YES	YES	YES	YES
modulo addressing	YES	YES	NO	NO
true circular addressing	NO	NO	NO	YES
delayed branch	NO	YES	YES	YES

(b)

Fig. 3.2: Hardware-supported DSP features
(a) Integer
(b) Floating point

• DSP56000 & DSP96000 (3.2)

- TMS320C25 & TMS320C50 (3.3)
- ADSP2100 & ADSP21020 (3.4)
- DSP32C & TMS320C30 (3.5)

The TMS320C25 is the very popular integer processor used for most of the code listings in this book. The TMS320C50 is a later version of the TMS320C25 with about double the machine cycle rate but source code compatible. Because it is the most recent of the designs, it contains all the features described in Chapter 2. The DSP96000 and ADSP21020 are floating point processors derived from the DSP56000 and ADSP2100 integer processors respectively. The DSP32 and TMS320C30 are floating point machines and claim to use a von Neumann architecture, with a single address space for program, data and stack. Note that the TMS320C30 is not related architecturally to the integer TMS320C25/C50. It will be remembered from Chapter 2 that we said that such an architecture cannot perform the multiply-accumulate operation efficiently and we shall explain how these 'von Neumann' DSPs function.

All the processors possess at least some of the special features described in general terms in Chapter 2 and these are summarised in Fig. 3.2.

There are some features shared by many of these DSPs which are not derived from theoretical concepts but result from practical considerations, such as the overriding need for high speed.

3.1.1   On-chip stack

It will be seen in Fig. 3.1 that six processors possess a small on-chip stack. This is used to support subroutine calls by holding the return address. As the stack uses a separate on-chip data bus, it means that a subroutine call can be as fast as a simple

processor	DSP56000	TMS320C25	TMS320C50	ADSP2100
nested hardware block repeat	YES	NO	NO	YES
host port (co-processor support)	YES	NO	NO *	YES
shadow registers	NO	NO	YES	YES
synchronous serial ports	YES	YES	YES	NO *

processor	DSP96000	DSP32C	TMS320C30	ADSP21020
nested hardware block repeat	YES	NO	NO	YES
host port (co-processor support)	YES	NO	NO	YES
shadow registers	NO	YES	NO	YES
synchronous serial ports	YES	YES	YES	NO*

*   But variants exist which have this capability

Fig. 3.3: Miscellaneous DSP features

branch or jump because the return address can be 'pushed' at the same time as the external data bus is being used to read machine code. However these small fast stacks are normally too limited in size to be used for the temporary saving of registers or the passing of parameters to a subroutine, which is common practice on von Neumann machines. The Motorola and Analog Devices processors also use the on-chip stack to save the essential parameters needed to support the block repeat (Section 2.6.2) without action by the programmer. This allows such loops to be nested (loop within a loop) in hardware without any software overhead.

3.1.2   Host port

DSPs are most often used in a conventional 'stand alone' manner in an embedded application where the program is stored in ROM or EPROM and runs from switch-on. An alternative scenario is where one or more DSPs are used as a co-processor to support a general purpose computer, for example in a personal computer. Several of the DSPs (Fig. 3.3) possess an i/o 'host port' designed for communication with a host processor and an on-chip bootstrap loader. In this case, the DSP subsystem can be ROMless and, at reset, the bootstrap loads a DSP program into its own local RAM from the host, via the host port. This is a particularly attractive option if the processor has a wide program memory data bus - the ADSP21020 bus has 48 bits and would require 6 external ROM devices.

3.1.3   Interrupts

Interrupts are often used to collect input data and deliver output data in any computer system. When a timer or an external device interrupts the processor, an interrupt subroutine handles the input/output. It has to save the processor status (any registers etc. used) first and restore it before returning to the main program. In a DSP, the high i/o sampling rate means that the operations to save and restore the registers used in the interrupt subroutine can use up a considerable proportion of the total processing power. The situation is made worse if the DSP has a stack which is too small to push the registers onto anyway. Some of the DSPs described here use an alternate set of 'shadow registers' where a copy of the main processor registers can be saved very quickly in order to support a single level but efficient interrupt.

A DSP interrupt subroutine is often very simple; perhaps just one instruction to read an input sample. It may therefore be practical to allocate some processor registers permanently for this purpose and avoid saving and restoring registers.

3.1.4   Synchronous serial ports

At least one fast (ca. 10 Mbits/s) synchronous serial port is a feature of most modern DSPs. The protocol employed is standard, being derived from serial buses used by the telecommunications industry and quite incompatible with the more familiar RS232 type of asynchronous port found on every personal computer. The

DSP serial port is extremely convenient for the interfacing of ADCs and DACs, especially as there are several devices that require no additional hardware for the interface. Several CODECs (coder-decoder) with combined ADC and DAC facilities are available at very low cost. A further incentive for using this serial port is that it avoids the practical problems of constructing interfaces to the data and address buses, which have to run at extremely high speed.

3.2   MOTOROLA DSP56000 & DSP96000 [19], [20]

The Motorola 96000 is a floating point DSP, whose architecture was derived directly from the earlier DSP56000 , which we shall therefore describe first.

3.2.1   Arithmetic on the DSP56000

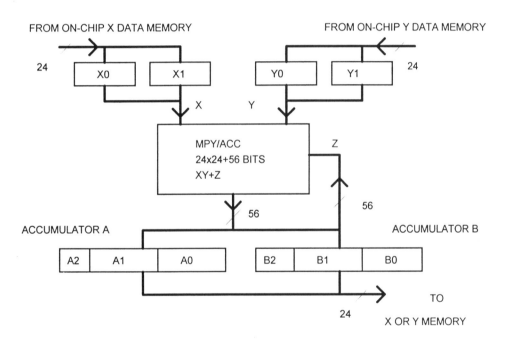

Fig. 3.4: DSP56000 arithmetic/logic unit

The Motorola DSP56000 (56000) is basically a 24 bit integer processor and has an arithmetic logic unit (ALU) which can perform a 24 x 24 multiply *and accumulate* (i.e. xy + z) in one machine cycle (better than 100 ns). As shown in Fig. 3.4, there are two pairs of 24 bit registers which can hold the input data for the ALU, X0, X1 and Y0, Y1. Two accumulators of 56 bits hold the result of the multiply/accumulate operation. A signed 24 bit multiply produces a 47 bit result assuming twos complement arithmetic with a range of -1 to +.9999 (see Section 2.2.3). Therefore the fact that there are an extra 9 accumulator bits means that 512 product accumulations can be calculated without any danger of accumulator overflow. The X and the Y registers used to supply ALU data are connected to two separate areas of on-chip memory, X memory and Y memory, by two 24 bit buses. Thus two operands can be delivered simultaneously to the ALU. Two pointers can be modified (e.g. post-increment) while an instruction is being executed. Processor instructions are stored in a third area of memory with its own 24 bit address and 24 bit data bus. This non von Neumann architecture obviously gives an improvement in the multiply-accumulate performance, because of its use of three parallel data buses. Machine code can be fetched at the same time as two data values are delivered to the multiplier. Note however that there is some loss of generality. In order to get this performance, the two streams of operands must be stored in separate memory spaces; this may not be convenient. (Consider for example accumulating the squares of any array. All the processors mentioned here have to use a dedicated instruction to square a value). This architecture is also rather algorithm specific. It is enlightening to consider how it would perform if the product of three rather than two values had to be computed. The X0, X1 and Y0,Y1 registers can be loaded with two new operands in the same machine cycle that multiplies the current register contents so that the multiply- accumulate computation rate can be maintained for a sequence of operations.

The 56000 uses an entirely separate off-chip memory space to hold the program using a third bus. All of the integer DSPs described in this chapter use a program memory area separate from that used for data. This is often known as the Harvard architecture. The idea of the Harvard architecture is not new; in fact it predates the single memory von Neumann architecture. It has the advantage that, at least conceptually, an instruction can be fetched on the program memory bus at the same time as data is transferred on a separate bus.

There is one important restriction in the use of the multiple buses of the 56000. The i.c. package only possesses a single external address/data bus. Therefore the advantages of the parallelism can only be obtained if memory on the processor chip can be used. In this case, 256 words of X and 256 words of Y memory are provided on chip. This might seem rather small in the context of a lot of general purpose algorithms but is quite substantial in terms of expected values of N in Eq. 2.1.

3.2.2   DSP56000 addressing modes

The 56000 uses 8 address pointer registers R0-R7, shown in Fig. 3.5. Each can point to X or Y or Program memory. Each address pointer has an associated offset register N0-N7 which holds the value to be added to the address pointer in post-increment addressing. There are also eight associated modifier registers, M0-M7 which control pointer arithmetic, e.g. modulo and bit reverse. Many instructions allow the programmer to specify two data move operations in parallel with the instruction. For example, the basic core of a multiply/accumulate operation might be:

```
REP    #K      ;repeat following instruction
MAC    X0,Y0,A    X:(R0)+,X0  Y:(R4)+,Y0
```

The first instruction repeats the multiply-accumulate that follows, see 2.6.1. The multiply-accumulate instruction MAC performs the following operations repeatedly:

$$A + X0 * Y0 \rightarrow A$$
$$X \text{ data at address } R0 \rightarrow X0$$
$$R0 + 1 \rightarrow R0$$
$$Y \text{ data at address } R4 \rightarrow Y0$$
$$R4 + 1 \rightarrow R4$$

Registers R0, R4 'walk through' memory holding lists of coefficients. Modulo addressing is used to create a circular buffer, so that the two instructions can perform the entire *for* loop shown in Fig. 2.11b at a rate approaching one multiply-accumulate per machine cycle.

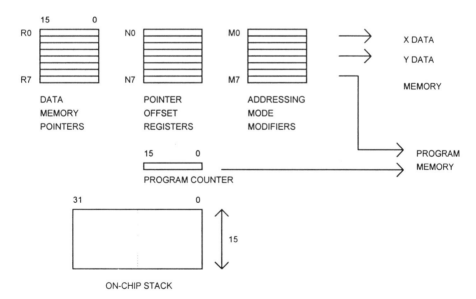

Fig. 3.5: DSP56000 address pointers

3.2.3   Motorola DSP96000

The 96000 is a powerful floating point processor that has complete hardware support for handling 32 bit floating point numbers. However its architecture is based on the 56000 already described. It has 512 words (32 bits) of X and Y memory on the chip. Floating point arithmetic is carried out in ten general purpose 96 bit registers, instead of the two accumulators used by the 56000. There are again 8 address pointer registers (of 32 bits), each with an offset register and a modifier register. As with the 56000, two address pointers can be used and modified by a single instruction, concurrently with its primary function. Note however that all three memory areas are enormously larger than those of its integer counterpart at 4 Gwords.

Many instructions can specify three operands, two sources and a destination. A 96000 instruction to multiply and accumulate is shown in Fig. 3.6a. In common with other modern processors the concurrent nature of the instruction is specified clearly by the source code. The concurrency is specified by putting the two components of the instruction on one line of the source; if they occupied two lines, two instructions would be generated and execution would take twice as long.

	INSTRUCTION	ACTION
(a)	FMPY S1,S2,D1 FADD.S S3,D2	D1 = S1 * S2; D2 = D2 + S3
(b)	FMPY S1 S2 D1 FADDSUB.S D3 D2	D1 = S1 * S2; D2 = D2 + D3; D3 = D3 - D2

Fig. 3.6:
(a) Multiply-accumulate on the DSP96000
(b) Highly parallel DSP96000 instruction

The extreme parallelism, which is found increasingly in modern DSPs can easily get very specialised and difficult to use except in a specific context. An example is shown in Fig. 3.6b. This is an inevitable part of the speed/generality trade-off inherent in the DSP philosophy.

Two full external data and address buses are available and can be used as host ports in coprocessor applications.

3.2.4   DSP56000 and DSP96000 Interrupts

In recognition of the need in DSPs for a simple but fast interrupt, the 56000 and 96000 DSPs support a special fast interrupt scheme. This executes two instructions specified by the programmer. No status is saved and so registers used by the main program must not be corrupted.

3.3   TMS320C25 AND TMS320C50 [29] [30]

The Texas Instruments TMS320C25 and its recent derivative, the TMS320C50 (C25/C50) are 16 bit integer DSPs. The C50 is source code compatible with the C25 and has a few extra facilities. However, its main virtue is that it runs at around twice the speed of the earlier processor, performing a machine cycle every 2, rather than every 4 clock cycles. We shall primarily discuss the C25, but except where indicated, everything that follows applies to both processors.

3.3.1   TMS320C25/C50 programmer's model

The C25 uses just two separate memory spaces, program memory and data memory, but has just a single external data/address bus. It does however have an amount of data memory on the chip with a second on-chip data bus. The main processor registers are shown in Fig. 3.7. Addition and subtraction is normally carried out in the single 32 bit accumulator and a 32 bit product register (P) holds the result of a 16 x 16 bit multiplication. Eight address pointer registers AR0 - AR7 provide indirect addressing of the *data* memory area only. Post-increment and post-decrement addressing of a single pointer by 1 or by the contents of an index register INDX (in the C25, INDX is AR0). are provided as well as bit reversed post-increment; examples are shown in Fig. 3.8. The TMS320C50 also provides circular addressing. At first glance, the C25 appears to be capable of incrementing just a single address pointer in one instruction, accessing its single data memory area. We shall therefore have to look more closely to see how it is possible for it to access the two lists of values during convolution computations.

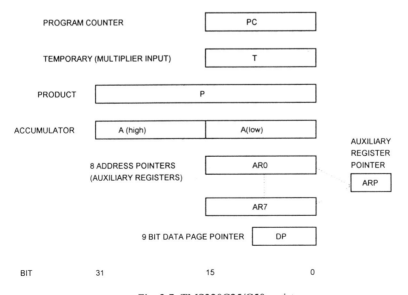

Fig. 3.7: TMS320C25/C50 registers

INSTRUCTION	ADDRESSING MODE ACTION
LAC *+	post-increment ARn by 1
LAC *0+	post-increment ARn by value in INDX
LAC *BR0+	post-increment ARn by INDX with reverse carry

Fig. 3.8: TMS320C25/C50 instructions to load accumulator
from data memory using indirect addressing modes

3.3.2   Multiply-accumulate on the TMS320C25/C50

In order to perform the convolution of Eq. (2.1) on the C25/C50, coefficients h(0) to h(N-1) are stored in consecutive *program memory* addresses and data x(n) to x(N-1) in on-chip data memory. The convolution is performed by the repetition of a specialised instruction MACD, e.g.

            RPTK N-1   ;repeat MACD N times
            MACD h0,*-   ;multiply-accumulate

Before execution, the current auxiliary register (ARn) is set up to point to the oldest data sample, x(N-1). MACD loads an internal pointer (ip) with the address of coefficient h(0) in program memory and at each repetition decrements ARn and increments ip by 1. The x( ) and h( ) data samples can be read simultaneously on the data and program memory data buses. Because MACD is repeated, it only has to be fetched once at the start of the convolution (Section 2.6.1) and instruction fetches do not occupy the program memory data bus.

Unlike the DSP56000, the ALU cannot compute XY + Z in one cycle, the multiplier is separate from the adder. The load, multiply and add operations are pipelined as shown in Fig. 3.9.

MACHINE CYCLE	T REG LOAD	MULTIPLY (new P)	ACCUMULATE (old P)	MOVE x( )
1	T=h(0)			
2	T=h(1)	P=T*x(N-1)		x(N)=x(N-1)
3	T=h(2)	P=T*x(N-2)	A=A+P	x(N-1)=x(N-2)
4	T=h(3)	P=T*x(N-3)	A=A+P	x(N-2)=x(N-3)
5				
	- - - etc. - - -			
N	T=h(N-1)	P=T*x(n-1)	A=A+P	x(n-2)=x(n-1)
N+1		P=T*x(n)	A=A+P	x(n-1)=x(n)
N+2			A=A+P	

Fig. 3.9: TMS320C25/C50 pipelined multiply-accumulate

In the first machine cycle, the T register is loaded with a coefficient from program memory. In the second, another coefficient is loaded while the first is multiplied by a sample value from data memory. In the third cycle the first product is being accumulated, see Fig. 3.9. In subsequent cycles, the pipeline is full and maximum speed is attained with the loading of two operands, a multiplication and an addition per cycle. At each iteration of the MACD instruction, the x sample read from data memory for the multiply is written to the adjacent (older) data location. This means that, when the iteration is complete, the x samples have all been shifted into the next 'older' location ready for reading another sample, as was shown in Fig. 2.8a. This unusual read plus write in a single machine cycle is only possible with on-chip memory.

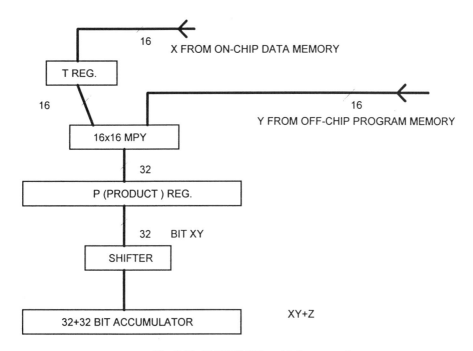

Fig. 3.10: TMS320C25 multiply-accumulate

The one remaining question concerns the avoidance of overflow; the accumulator can only store 32 bits. As shown in Fig. 3.10, TMS320 has a shifter between the product register and the accumulator. To avoid overflow, the product can be shifted right by six bits before the accumulation takes place. Of course some bits of low significance will be lost. This does not matter because the 32 bit result will normally be truncated to 16 bits anyway. Examples of TMS320C25/C50 convolution will be found in Chapter 6 on FIR filters.

3.3.3   C25/C50 direct addressing

In the C25 and C50, as with all DSPs, the designers have been at pains to avoid the speed penalty normally associated with absolute addressing. In absolute addressing, the complete 16 bit address would occupy a whole 16 bit word of program memory. This means that a second word would have to be used to specify the instruction which would therefore need two machine cycles to execute. The unusual solution here has been to divide data memory into 512 pages, each of 128 words. A 9 bit data page pointer register is loaded by the programmer to specify the page in use. Thus the instruction needs only to specify the 7 low address bits and can fit in a single word of program memory and execute in one machine cycle. Of course this does not affect indirect addressing as the address pointers AR0-AR7 can hold all 16 data memory address bits.

3.3.4   C25/C50 Interrupts

The C50 supports a fast interrupt by providing a set of shadow registers for saving the main program status. However in both the C25 and the C50 it is often practical to allocate an address pointer register solely to the interrupt subroutine. A useful feature of the instruction set is that data may be transferred by IN or OUT instructions directly between the port and data memory, rather than via the accumulator. It may therefore be possible to avoid the overhead of saving the accumulator in an interrupt subroutine.

3.4   ADSP2100 AND ADSP21020 [1] [2]

The Analog Devices DSPs are among the most recent of the DSP designs. Unlike the TMS320C50, the designers did not have to retain compatibility with previous products. As a consequence, the designers have incorporated most of the best features of other DSP designs. These are two unusual, advanced and powerful processors.

3.4.1   ADSP2100 and ADSP21020 memory

It may have been noticed from Fig. 3.1 that the Analog Devices ADSP2100 and its floating point derivative the ADSP21020 have no on-chip data or program memory - the only DSPs in our selection that are like this. The reason for this is that modern packaging technology has allowed the designers to provide two complete external address/data/control buses, one for a program memory area and one for data memory. These are used for the simultaneous transfer of two operands to the multiplier which, like the TMS320C25/C50 case are held in pairs in program and data memory. In addition to a single instruction repeat facility, these two processors have another method for avoiding instruction fetches. An on-chip cache memory is provided. When an instruction is fetched and executed, a copy is stored in the cache. In a small iterated block of code, when the second and subsequent repetitions are performed, the instructions can be fetched from the cache, thus releasing both external buses for data transfers. This is obviously a more versatile scheme than the single instruction repeat idea and extends the range of situations where a single cycle multiplication can be performed.

The data and address bus sizes of these processors are summarised in Fig. 3.11.

PROCESSOR	PROGRAM MEMORY		DATA MEMORY	
	DATA BUS	ADDRESS BUS	DATA BUS	ADDRESS BUS
ADSP2100	24 BITS	14 BITS (16K)	16 BITS	14 BITS (16K)
ADSP21020	48 BITS	24 BITS (16M)	40 BITS	32 BITS (4G)

Fig. 3.11: ADSP2100/ADSP21020 address and data bus sizes

A very unusual feature is that the data buses are longer than the address buses. In fact in both cases, the program memory word length is considerably longer than the program memory address, the data memory address and the data memory word size. This means that any of these can be specified within a single word machine code instruction. In consequence a branch (jump) instruction, an absolute data memory reference or an 'immediate' data constant reference can occupy one word and execute in a single machine cycle. In fact every instruction executes in one machine cycle. The price of this is the very limited address spaces on the integer processor and the high cost of external memory on the floating point one. Versions with substantial on-chip program and data memory are now available with a single address/data external bus.

3.4.2   Miscellaneous ADSP2100 and ADSP21020 facilities

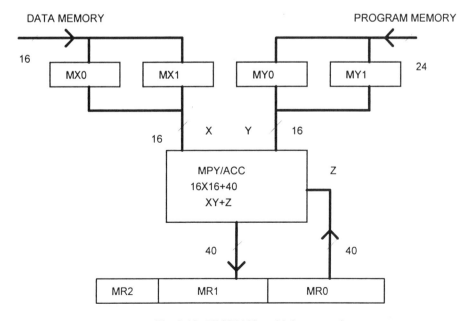

Fig. 3.12: ADSP2100 multiply-accumulate

These two processors possess pretty well all the desirable functions of other DSPs. The 2100 multiplier is shown in Fig. 3.12 and can be seen to be very like that of the DSP56000 shown in Fig. 3.4, except that one operand is supplied from program memory like the C25/C50. Two sets of address pointers can be modified in one machine cycle and support all the important addressing modes mentioned in Chapter 2. The reader is referred to the manufacturer's data books for further information.

3.5   DSP32 AND TMS320C30 [3] [31]

The AT&T DSP32 was easily the first floating point DSP to be manufactured around 1984 and claims to use a von Neumann architecture; i.e. there is only a single address space for both program and data. The Texas Instruments TMS320C30 is also floating point but was not derived from the AT&T product. Both use a 32 bit data bus to support something like the IEEE short floating point format and have a very large address space. As we have pointed out that a von Neumann system is incompatible with efficient DSP operation, some explanation is required.

Firstly, one might wonder why the von Neumann scheme has been preferred anyway, particularly by Texas Instruments who abandoned the C25/C50 Harvard architecture in the C30. A credible explanation is to assist the use of high level language, such as C, rather than assembler language. A conventional C programmer does not expect to have to specify the type of memory or location in memory that stores his data, indeed ANSI C does not anticipate a non-von Neumann architecture. In addition, the normal HLL vehicle for transferring data to and from a C function (subroutine) is the system stack. We have seen already that, for the sake of speed, most DSPs use an on-chip hardware stack which is too small for this purpose (note though that the normal DSP post-increment addressing mode allows a stack to be efficiently implemented in software).

3.5.1   Accessing the DSP32 memory space

These floating point processors perform several accesses *from the same logical memory space* per machine cycle. The DSP32 is able to perform four: instruction fetch, two data reads and a data write. These occur consecutively in a machine cycle in the order write, read 1, read 2, fetch. This sounds like the perfect scheme until one realises that it is only made possible by a pipelined design and the memory accesses pertain to successive instructions as shown in Fig. 3.13. There is a further restriction on memory accesses in the DSP32. Although there is only one logical address space, memory is divided physically into four blocks with separate buses; three are on-chip, each with two buses and the fourth is accessed by the single external bus. Every bus may be accessed twice per instruction cycle but, to achieve full speed, consecutive accesses must not be made to the same physical memory block. This is because, while data is being accessed, the next address is set up. If consecutive accesses are made, wait states are automatically inserted to slow down execution. Obviously this introduces potential penalties for program and data location which do not exist on a 'true' von Neumann machine and so the DSP32 programmer must be aware of physical addresses.

CYCLE	INSTRUCTION	READ	READ	WRITE
1	I(1)			
2	I(2)			
3	I(3)	X(1)	Y(1)	
4	I(4)	X(2)	Y(2)	
5	I(5)	X(3)	Y(3)	
6	I(6)	X(4)	Y(4)	Z(1)
7	I(7)	X(5)	Y(5)	Z(2)

Fig. 3.13: Pipelined memory accesses in the DSP32C
instruction I(n) performs $Z(n) = X(n)Y(n) + Z(n)$

As mentioned in Section 2.6.1, convolution on the DSP32 is carried out by the repeated operation of a single instruction. This is carried out as shown in Fig. 3.14. Compare this to Fig. 3.13.

CYCLE	INSTRUCTION	READ	READ	WRITE
1	fetch instruction			
2				
3		read *r4++	read *r2++	
4				
5				
6				write *r3++
7				

Fig. 3.14: Memory accesses in the DSP32 during
a1 = a1 + (*r3++ = *r4++) * *r2++

If the maximum speed is to be attained, consecutive memory accesses must be via different physical buses, so that for example the two data arrays should not be in the same physical block and the instructions should not be stored in the block that r3 points to. All this means that we are looking at the exception that proves the rule: the unrestricted consecutive accesses which are implied with a single linear memory space are not compatible with the parallel movements of data and instructions, necessary in the fastest DSP implementations.

3.5.2   The TMS320C30 memory space

The TMS320C30 is also quasi von Neumann and uses a similar system to the DSP32C, with four physical on-chip memory blocks (one an instruction cache), each of which can support two accesses per instruction cycle. In this case however, any two accesses may be carried out in a single machine cycle. This means that there may be two writes or two unrestricted reads (including instruction fetch) per cycle per physical memory block. However, exceeding this results in a delay of a complete machine cycle, instead of a quarter cycle as in the DSP32C. Therefore, in the TMS320C30 also, program and data location in physical memory is significant.

4

Discrete Fourier Transform : properties, uses, and computations

4.0   INTRODUCTION

We start with the discrete-time Fourier transform (DtFT) and develop from it the discrete Fourier transform (DFT). This is followed, in Sections 4.2 and 4.3, with DFT use in computations of the convolution and correlation sequences. In Section 4.4, the fast Fourier transform (FFT) is introduced as an efficient method for DFT computation. In the following two Sections, 4.5 and 4.6, the Goertzel algorithm and the chirp-z transform are introduced as alternative approaches of the DFT computation. In Section 4.7 some computer studies are presented, using FFT and Goertzel algorithm. Chapter 5 deals with DSP implementations of Fourier and Goertzel algorithms.

4.1   DISCRETE FOURIER TRANSFORM (DFT)

An analogue signal x(t) is described in the time-domain, and also in the frequency-domain in terms of its Fourier transform X(ω). In both cases, variables are continuous in time and frequency. For digital signal processing both x(t) and X(ω) need to be sampled into a finite set of values. This is firstly done by sampling in the time-domain at uniform intervals of time T, producing a discrete-time sequence x(nT) or simply written as x(n), where n is an integer. The sampling process, discussed in Section 1.1, produces a periodic spectrum at the sampling frequency and its multiples (Eq. 1.3).

The time/frequency relationship for a time-sampled signal are given by Eqs. (1.5) and (1.6), and rewritten here as

$$X(\theta) = \sum_{n=-\infty}^{\infty} x(n) e^{-jn\theta} \qquad (4.1)$$

$$x(n) = \frac{1}{2\pi} \int_{-\pi}^{\pi} X(\theta) e^{jn\theta} \, d\theta \qquad (4.2)$$

where $\theta = \omega T$. As mentioned in Section 1.1, this set of equations is often referred to as the discrete-time Fourier transform (DtFT). We note again that $x(n)$ are discrete-time signal samples, but $X(\theta)$ is a continuous function of frequency ($\theta = \omega T$), and hence it is not in a suitable computational form. Therefore, we consider next the representation of a sequence $x(n)$ by samples of its spectrum $X(\theta)$ which will lead to the discrete Fourier transform (DFT).

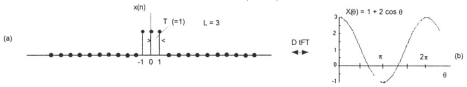

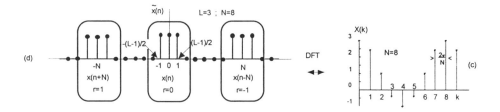

Figure 4.1:
  (a) Finite length sequence
  (b) Frequency spectrum of (a)
  (c) Sampled spectrum of (b)
  (d) Periodic time sequence for X(k)

For an illustration of this process, we take a finite length signal sequence $x(n)$ consisting of three samples, Fig. 4.1(a). Its spectrum, from Eq. (4.1) is given by $X(\theta) = 1 + 2 \cos \theta$, which is plotted in Fig. 4.1(b). In this case the time sampling interval T defines the frequency period of the sampling frequency

$$\omega_s = \frac{2\pi}{T} \, , \quad \text{or} \quad \theta_p = \omega_s T = 2\pi \, .$$

The spectrum is now sampled at interval $\theta_s = 2\pi/N,$ where N is a chosen number of samples, as shown in Fig.4.1(c). This sampling results in periodic repetition of x(n), denoted by $\tilde{x}(n)$ in Fig. 4.1(d), with the period given by

$$\frac{2\pi}{\theta_s} = N \quad .$$

It can be shown [27] that this periodic sequence is expressed as

$$\tilde{x}(n) = \sum_{r=-\infty}^{\infty} x(n + rN) \quad , \tag{4.3}$$

which is the counterpart to Eq. (1.3) in the frequency domain. The sequence $\tilde{x}(n)$ is called the periodic extension of x(n). We are free to choose N, i.e. the number of samples of $X(\theta)$ over the period $0 \leq \theta \leq 2\pi$. Since N is also the period of $\tilde{x}(n)$, we must not choose N too small. To illustrate this point let x(n) be a finite-duration sequence containing L samples.

As shown in Fig.4.2, if $L \leq N$, then

$$\tilde{x}(n) = \begin{cases} x(n) & \text{for } 0 \leq n \leq L - 1 \\ 0 & \text{for } L \leq n \leq N - 1 \end{cases} \quad . \tag{4.4}$$

This means that when $L \leq N$, the finite duration sequence x(n), for $0 \leq n \leq L-1$, can be recovered uniquely from the first period of $\tilde{x}(n)$. It is also seen that for $N > L$, there are zero points in each period of $\tilde{x}(n)$. If $L > N$, the duration of x(n) is longer than the period of $\tilde{x}(n)$, an overlap occurs causing an error when we try to retrieve x(n). In this case, time-aliasing is said to occur. To prevent time-aliasing, $X(\theta)$ must be sampled over $0 \leq \theta \leq 2\pi$ at least as many times as there are elements in x(n), or $N > L$ times.

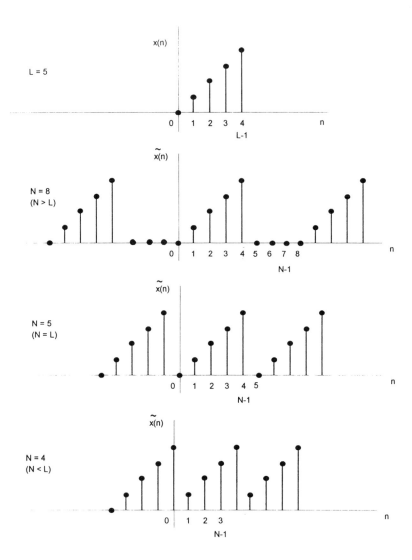

Figure 4.2: Illustration of time domain aliasing and its avoidance for N > L

We derive next the DFT from Eq. (4.1). As shown in Fig. 4.1, the discrete frequencies are given by $\theta_k = (2\pi/N)k$. Hence $X(\theta)$ in Eq. (4.1) becomes $X(k)$ given by

$$X(k) = \sum_{n=-(L-1)/2}^{(L-1)/2} x(n)\, e^{-j2\pi kn/N} \quad ,$$ (4.5)

where the limits are determined from the first period of $\tilde{x}(n)$ as indicated in Fig. 4.1(d). In actual processing we need, for counting processes in algorithms, the index to start at zero. This is accomplished by shifting the starting point of the data (L-1)/2 positions which changes the above expression into

$$X(k) = \sum_{n=0}^{L-1} x(n) e^{-j2\pi kn/N} \quad . \tag{4.6}$$

The (L-1)/2 shift, will affect only the phase angles of this transform [12]. The upper limit is normally taken for convenience as N-1, with x(n) = 0 for L ≤ n ≤ N-1. Therefore, **the N-point DFT** standard form is given by

$$X(k) = \sum_{n=0}^{N-1} x(n) e^{-j2\pi kn/N}, \qquad 0 \le k \le N\text{-}1 \quad . \tag{4.7}$$

The inverse discrete Fourier transform (**IDFT**) is given by

$$x(n) = \frac{1}{N} \sum_{k=0}^{N-1} X(k) e^{j2\pi kn/N}, \qquad 0 \le n \le N\text{-}1 \tag{4.8}$$

One can obtain the above in a non-rigorous way from Eq. (4.2) by setting $d\theta = 2\pi/N$, changing the integral to the summation, and changing limits from ±(N-1)/2 to 0 and N-1 similarly to derivations of Eq. (4.7). It has been seen earlier, for Eq. (4.7), that when L<N we use (N-1-L) zeros for x(n). This is referred to as zero-padding, producing a more detailed X(k).

For infinite-duration sequences, there is always some amount of time-aliasing when we use the DFT relationships. The value of N is then chosen to be large enough to make this error negligible in a particular application.

Example 4.1

The familiar infinite-duration impulse response of the first order IIR filter, is given by $x(n)=a^n$, n ≥ 0, and 0 < a < 1. The corresponding spectrum is sampled at intervals $2\pi/N$, and we want to find the effect of aliasing into basic range 0 ≤ n ≤ N-1. Sketching the periodic extension, Eq. (4.3) for this case, one concludes that only r = 0 to ∞ (i.e. the l.h.s.) can cause error in the basic range. Hence, Eq. (4.3) gives

$$\tilde{x}(n) = \sum_{r=0}^{\infty} a^{n+rN} = a^n \sum_{r=0}^{\infty} \left(a^N\right)^r$$

$$\therefore \tilde{x}(n) = \frac{a^n}{1-a^N}, 0 \le n \le N-1 \quad .$$

This result shows that $\tilde{x}(n) = a^n = x(n)$, for $0 < a < 1$ and $N \to \infty$.

Example 4.2

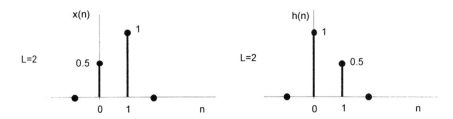

Figure 4.3: Two finite length sequences for Example 4.2

Two finite duration sequences (L=2) are given in Fig. 4.3. Calculate their spectra using DFT, Eq. (4.7), with N=L=2.

Solution
For N=2, Eq. (4.7) reduces to

$$X(k) = x(0) + x(1)\, e^{-j\pi k}, \text{ k=0, 1}$$
$$X(0) = x(0) + x(1) = 1.5 \text{ and } X(1) = x(0) - x(1) = -0.5,$$

Similarly for H(k) we obtain,

$$H(0) = 1.5 \text{ and } H(1) = 0.5.$$

4.2 USE OF THE DFT IN CONVOLUTION

The convolution operation, Eq. (1.29), often denoted by y(n) = h(n) * x(n), is rewritten here for N points as

$$y(n) = \sum_{m=0}^{N-1} h(m) \cdot x(n-m) \quad , \tag{4.9}$$

where h(m) is a finite N-length impulse response sequence of a linear network. The input x(n) and output y(n) sequences can be of finite or infinite length.

The discrete-time Fourier transform (DtFT) converts the convolution of $h(n)$ and $x(n)$ into multiplication of $H(\omega)$ and $X(\omega)$:

$$y(n) = h(n) * x(n) \xleftrightarrow{\text{DtFT}} Y(\omega) = H(\omega) \cdot X(\omega) \quad . \tag{4.10}$$

For computer applications we need to use the DFT so we have [27]:

$$y(n) = h(n) \, (*) \, x(n) \xleftrightarrow{\text{DFT}} Y(k) = H(k) \cdot X(k) \quad . \tag{4.11}$$

Using then the IDFT we obtain $y(n)$. This method, known as fast convolution, seems cumbersome, but it is often faster than the time domain convolution, mainly due to the speed of the FFT (Section 4.4). However, in the DFT relations all sequences are periodic with period N. The convolution, in this case, is called **circular convolution**, denoted by

$$y(n) = h(n) \, (*) \, x(n), \qquad \text{and given by}$$

$$\tilde{y}(n) = \sum_{m=0}^{N-1} \tilde{h}(m) \cdot \tilde{x}(n-m, \text{mod} N), \quad 0 \le n \le N-1 \quad , \tag{4.12}$$

while the convolution given by Eq. (4.9) is called **linear convolution**. In Eq. (4.12), $\tilde{x}(n-m)$ involves time values outside the range $0 \le n < N$. To keep in the range we use mod N notation, which means $(p, \text{mod } N) = p + qN$ where integer q is such that $0 \le (p, \text{mod } N) < N$; this is illustrated in the example given below with $N = 3$.

The $\tilde{y}(n)$ sequence can be obtained graphically or algebraically. The multiplications of $\tilde{h}(n)$ and $\tilde{x}(n-m)$, for the graphical method are done over the window $m = 0, 1, ..., N-1$, and summed to obtain $\tilde{y}(n)$ within the window $n = 0, 1, ..., N-1$. For the algebraic method we expand Eq. (4.12), for say N=3, as follows

$$\tilde{y}(n) = \sum_{m=0}^{2} \tilde{h}(m) \, \tilde{x}(n-m, \text{mod } 3), \quad n = 0,1,2$$

$$\therefore \quad \begin{aligned} y(0) &= h(0)x(0) + h(1)x(-1) + h(2)x(-2) \\ y(1) &= h(0)x(1) + h(1)x(0) + h(2)x(-1) \\ y(2) &= h(0)x(2) + h(1)x(1) + h(2)x(0) \quad . \end{aligned}$$

Since x(-1) and x(-2) are outside the range $0 \le n \le 2$ we use q=1, to move them into the range by adding N=3 to each index and hence x(-1, mod 3) = x(2), x(-2, mod 3) = x(1). Various aspects of the linear and circular convolution are shown through the following Examples 4.3, 4.4 and 4.5.

Example 4.3

We use the two finite duration sequences defined in Example 4.2 and form the linear and circular convolutions as illustrated in Fig.4.4, and Fig. 4.5.

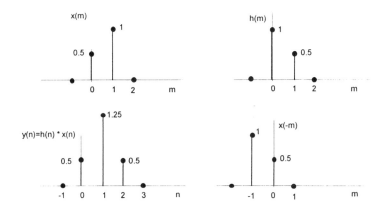

Figure 4.4: Linear convolution using Eq. (4.9)

In Fig. 4.4 we have the result of linear convolution obtained by applying Eq.(4.9). The sequence x(n-m) is obtained from x(m) by, first folding x(m) about m=0, resulting in x(-m). It is then multiplied by h(m), producing y(0). Then, the folded sequence is shifted by n to yield x(n-m), multiplied by h(m) and summed for y(n). If we want to apply the DFT to perform convolution, all the sequences involved become periodic (due to sampling in the frequency domain). This is shown in Fig. 4.5 where we have assumed a 2-point DFT (N=2). Performing now the circular convolution, Eq. (4.12), we obtain the periodic sequence $y(\tilde{n})$. It may seem initially that N=2 would be sufficient to avoid the aliasing since L=2, but the result for $\tilde{y}(n)$ = {1, 1.25} is different from the desired linear convolution result y(n) = {0.5, 1.25, 0.5}. The reason for this is discussed below in Example 4.4

Example 4.4

It is convenient here to show the application of the inverse DFT to the case in Fig. 4.5.

In example 4.2 we have derived the DFT for sequences used in Fig. 4.4. From these results we calculate

$$Y(k) = H(k) X(k) , k = 0,1 (N=2)$$

and obtain $Y(0) = 2.25, Y(1) =- 0.25.$

Applying now the inverse DFT to this case, we have

$$y(n) = \frac{1}{2} \sum_{k=0}^{1} Y(k) \, e^{j\pi kn} \quad n = 0,1$$

from which we obtain y(0) = 1, and y(1) = 1.25 in agreement with the result shown in Fig. 4.5.

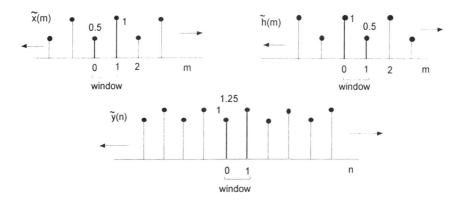

Figure 4.5: Circular convolution, Eq. (4.11) with N = 2

The DFT application in Examples 4.3 and 4.4 showed that for $N = L = 2$ time-aliasing occurs producing the wrong result. For the fast convolution we apply the inverse DFT which must contain the number of points in y(n). This number for the linear convolution is given by $N_y = N_x + N_h - 1$. In our example this means $N_y = 3$, also seen in the linear convolution result, Fig. 4.4.

Now, with $N_y = N = 3$, our periodic extension signals are shown in Fig. 4.6, and the result agrees with the linear convolution in Fig. 4.4.

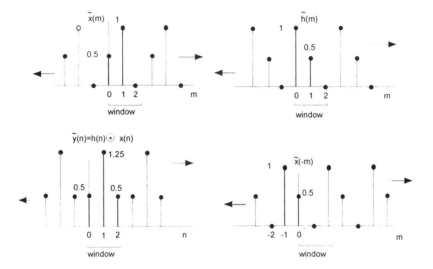

Figure 4.6: As in fig 4.5, but N = 3

Example 4.5

This is similar to Example 4.4 but now using a 3-point DFT. The input sequence DFT is given by

$$X(k) = \sum_{n=0}^{2} x(n) e^{-j2\pi kn/3} \qquad , k = 0,1,2 \quad ,$$

which produces $X(0) = 1.5$

$X(1) = 0.5 + e^{-j2\pi/3}$

$X(2) = 0.5 + e^{-j4\pi/3}$.

Similarly for $\{h(n)\}$ we obtain

$H(0) = 1.5$

$H(1) = 1 + 0.5 \, e^{-j2\pi/3}$

$H(2) = 1 + 0.5 \, e^{-j4\pi/3}$.

The output spectral components: $Y(k) = H(k) X(k)$, $k = 0$, 1, 2, are obtained as

$Y(0) = 2.25$

$Y(1) = 0.5 + 1.25 \, e^{-j2\pi/3} + 0.5 \, e^{-j4\pi/3}$

$Y(2) = 0.5 + 1.25 \, e^{-j4\pi/3} + 0.5 \, e^{-j8\pi/3}$

The required $\{y(n)\}$ sequence is then calculated from the inverse IDFT

$$y(n) = \frac{1}{3} \sum_{k=0}^{2} Y(k) e^{j2\pi kn/3}, n = 0,1,2.$$

For n=0

$$y(0) = \frac{1}{3}\left(3.25 + 1.25 e^{-j2\pi/3} + 1.75 e^{-j4\pi/3} + 0.5 e^{-j8\pi/3}\right) = 0.5 ,$$

and similarly for n = 1, n = 2, we would obtain y(1) = 1.25 and y(2) = 0.5 in agreement with the result in Fig. 4.6.

The calculations performed in Example 4.5 show that even a 3-point DFT is quite tedious to use. However, its advantage is that the computations can be made fast by using computers with a suitable algorithm. One of these algorithms, the fast Fourier transform (FFT), is explained in Section 4.4.

4.3 USE OF THE DFT IN CORRELATION

The convolution Eq. (4.9) is a special case which applies to processing of the input signal x(n) through a causal network with a finite length impulse response h(n). In general the convolution equation for two sequences u(n) and v(n), is given by

$$c_{uv}(n) = \sum_{m=-\infty}^{\infty} u(m) v(n-m) \tag{4.13}$$

Correlation is an operation that closely resembles convolution. In both cases, two signal sequences are involved. While the convolution gives the output of a linear processor such as a filter, the correlation is a measure of similarity between two signals. Correlation of signals is often encountered in radar, sonar, digital communications and other areas of science and engineering.

For two signal sequences u(n) and v(n), the cross-correlation of u(n) and v(n) is a sequence $r_{uv}(n)$ defined as

$$r_{uv}(n) = \sum_{m=-\infty}^{\infty} u(m) v(m-n), n = 0, \pm 1, \pm 2, \ldots \tag{4.14}$$

The index n is the time shift parameter. The order of subscripts, with u preceding v, indicates the direction in which one sequence is shifted relative to the other. In Eq. (4.14) the sequence u(m) is unshifted and v(m) is shifted by n units of time, to the right for n positive and to the left for n negative.

Example 4.6

Two sequences are given below

$$u(m) = \left\{ ...0,1,\underset{\uparrow}{2}, 1,-1,0,... \right\}$$

$$v(m) = \left\{ ...,0,\underset{\uparrow}{1}, 2,3,1,0.. \right\} \quad ,$$

where the arrows denote m=o. The result for the cross-correlation is

$$r_{uv}(n) = \left\{ ...0,1,5,9,7, \underset{\substack{\uparrow \\ n=0}}{1} ,-1,-1,0,.. \right\} \quad .$$

As in the case of the convolution operation, the correlation operation for short duration sequences is easy to compute by means of graphs or using a line presentation of sequences as in Example 4.6. For more complicated and long sequences it is better to use a computer program. If there is a convolution program available one can use it by folding $v(m)$ to $v(-m)$, or one can have a specific program for the cross-correlation, [27]. The correlation computation can be further simplified, similarly to the convolution, by working in the frequency domain.

As before, care must be taken in choosing the number of samples N in the frequency domain. This is to avoid aliasing due to the circular correlation $\tilde{r}_{uv}(n)$, so that it gives the correct result for $r_{uv}(n)$ in the interval $0 \le n \le N-1$.

In the special case of $v(m) = u(m)$, we obtain the auto-correlation of $u(m)$. It should also be mentioned that when dealing with finite-duration sequences, it is usual to express the auto-correlation and cross-correlation in terms of finite limits on the summation.

4.4 CALCULATION OF THE DFT : FFT ALGORITHMS

We rewrite the DFT Eq. (4.7) as

$$X(k) = \sum_{n=0}^{N-1} x(n) W_N^{kn}, \quad 0 \le k \le N - 1, \tag{4.15}$$

where $W_N = e^{-j2\pi/N}$. Direct computation of the DFT is inefficient because it does not exploit the symmetry and periodicity properties of the phase factor W_N.

The computationally efficient algorithms for the DFT are developed by using an index mapping (a change of variables). For this purpose we assume that the sequence length can be expressed as a product of two factors, that is

$$N = N_1.N_2 \ . \tag{4.16}$$

If N is a prime number, we can pad any sequence with zeros to satisfy a factorisation of the form (4.16). Now the sequence $x(n)$, $0 \leq n \leq N - 1$, can be stored in either a one-dimensional array indexed by n or a two-dimensional array indexed by n_1 and n_2, where $0 \leq n_1 \leq N_1 - 1$, and $0 \leq n_2 \leq N_2 - 1$ (note that we shall take n_1 to be the column index and n_2 to be the row index). A similar indexing is arranged for k. For the following development we represent the indices n and k as

$$n = N_2 \, n_1 + n_2 \quad , \tag{4.17a}$$

where $0 \leq n_1 \leq N_1 - 1$ and $0 \leq n_2 \leq N_2 - 1$,

$$k = k_1 + N_1 \, k_2 \quad , \tag{4.17b}$$

where $0 \leq k_1 \leq N_1 - 1$ and $0 \leq k_2 \leq N_2 - 1$.

(The above index mappings are a special case of general mappings, examined by Burrus and Parks [5], which are unique (one-to-one) with uncoupled variables).

     As n_1 and n_2 take on all possible values in the indicated ranges, n goes through all possible values from 0 to $(N - 1)$ with no values repeated. This is also true for the frequency index k. Using the mappings (4.17) in Eq. (4.15) we obtain

$$X(k) = X(k_1, k_2)$$

$$= \sum_{n_2=0}^{N_2-1} \sum_{n_1=0}^{N_1-1} x(n_1, n_2) W_N^{kn} \quad ,$$

where $kn = (k_1 + N_1 k_2)(N_2 n_1 + n_2) = N_2 k_1 n_1 + k_1 n_2 + N_1 N_2 k_2 n_1 + N_1 k_2 n_2$

$$\therefore X(k_1, k_2) = \sum_{n_2=0}^{N_2-1} \sum_{n_1=0}^{N_1-1} x(n_1, n_2) W_N^{N_2 k_1 n_1} W_N^{k_1 n_2} W_N^{N_1 N_2 k_2 n_1} W_N^{N_1 k_2 n_2} \ .$$

Since

$$W_N^{N_2 k_1 n_1} = W_{N_1}^{k_1 n_1}, W_N^{N_1 k_2 n_2} = W_{N_2}^{k_2 n_2}, \quad \text{and} \quad W_N^{N_1 N_2 k_2 n_1} = 1,$$

we have

$$X(k_1, k_2) = \sum_{n_2=0}^{N_2-1} \left[\left(\sum_{n_1=0}^{N_1-1} x(N_2 n_1 + n_2) W_{N_1}^{k_1 n_1} \right) W_N^{k_1 n_2} \right] W_{N_2}^{k_2 n_2}, \qquad (4.18)$$

where $0 \le k_1 \le N_1 - 1$, and $0 \le k_2 \le N_2 - 1$. The inner sum is the set of N_1 - point DFTs of the N_2 rows, which we denote as follows

$$f(n_2, k_1) = \sum_{n_1=0}^{N_1-1} x(N_2 n_1 + n_2) W_{N_1}^{k_1 n_1}, \qquad (4.19)$$

where $0 \le k_1 \le N_1 - 1$ and $0 \le n_2 \le N_2 - 1$. Since the input sequence is not needed again, the N_1 - point row DFTs can be stored in the same array locations as the original signal samples. This procedure is referred to as in-place calculation. The next step is the multiplication of row DFTs by the factors $W_N^{k_1 n_2}$ to obtain

$$g(n_2, k_1) = f(n_2, k_1) W_N^{k_1 n_2}, \qquad (4.20)$$

where $0 \le k_1 \le N_1 - 1$ and $0 \le n_2 \le N_2 - 1$. This operation can also be done in-place. The factors $W_N^{k_1 n_2}$ are called the twiddle factors. Without these factors expression (4.18) would be a two-dimensional DFT.

Finally, the outer sum in Eq. (4.18) represents the set of N_2-point DFTs of the columns of the array given by

$$X(k_1 + N_1 k_2) = \sum_{n_2=0}^{N_2-1} g(n_2, k_1) W_{N_2}^{k_2 n_2}, \qquad (4.21)$$

where $0 \leq k_1 \leq N_1 - 1$ and $0 \leq k_2 \leq N_2 - 1$. The column DFTs can be stored in the same locations as the input column $g(n_2, k_1)$. Therefore, the whole computation can be done completely in-place. The input is entered into the array according to the index map of Eq. (4.17a), the output DFT values must be extracted from the array according to Eq. (4.17b).

4.4.1 FFT algorithm, radix 2, decimation-in-frequency

The most common and most efficient form of the FFT uses all dimensions of the same length. This length is called the radix of the algorithm, denoted by R. The DFT of length N is then expressed as

$$N = R^M \quad , \qquad (4.22)$$

which gives M dimensions, each of length R. In the following example we develop the decimation-in-frequency FFT algorithm for the case of $R = 2$, i.e $N = 2^M$. With reference to Eq. (4.16), we choose for this case $N_1 = 2$ and $N_2 = N/2$. From Eq. (4.18) it follows that

$$X(k_1 + 2k_2) = \sum_{n_2=0}^{(N/2)-1} \left[\left(\sum_{n_1=0}^{1} x[(N/2)n_1 + n_2] W_2^{k_1 n_1} \right) W_N^{k_1 n_2} \right] W_{N/2}^{k_2 n_2} \qquad (4.23)$$

for $0 \leq k_1 \leq 1$ and $0 \leq k_2 \leq (N/2) - 1$.

The two-dimensional array representation of the input is shown in Fig. 4.7(a). The first column ($n_1 = 0$) of the array is the first half of the input sequence, and the second column ($n_1 = 1$) is the second half of the input sequence. The 2-point DFTs, in the inner sum of Eq. (4.23), are given by

$$f(n_2, k_1) = \sum_{n_1=0}^{1} x[(N/2)n_1 + n_2] W_2^{k_1 n_1} = x(n_2) + (-1)^{k_1} x[(N/2) + n_2] \qquad (4.24)$$

with $0 \leq k_1 \leq 1$, and $0 \leq n_2 \leq (N/2) - 1$. (Note : $W_2 = e^{-j\pi} = -1$).
The result is shown in the two-dimensional array Fig. 4.7(b). After multiplying by the twiddle factors we obtain the new array shown in Fig. 4.7(c), where in general

$$g(n_2, k_1) = f(n_2, k_1) W_N^{k_1 n_2} \quad , \qquad (4.25)$$

with $0 \leq k_1 \leq 1$ and $0 \leq n_2 \leq (N/2) - 1$.
Finally, the (N/2)-point DFT is given by the outer sum over the columns n_2 :

$$X\left(k_1 + 2k_2\right) = \sum_{n_2=0}^{(N/2)-1} g\left(n_2, k_1\right) W_{N/2}^{k_2 n_2} \quad , \tag{4.26}$$

for $0 \le k_1 \le 1$, and $0 \le k_2 \le (N/2) - 1$. This results in the two dimensional array in Fig. 4.7(d).

x(n) = x[(N/2) * n1 + n2]

n2	n1 = 0	n1 = 1	
0	x(0)	x(N/2)	
1	x(1)	x[(N/2) + 1]	
*	*	*	
*	*	*	(a)
*	*	*	
(N/2)-1	x[(N/2)-1]	x(N-1)	

f(n2,k1)

n2	k1=0	k1=1	
0	f(0,0)=x(0)+x(N/2)	f(0,1)=x(0)-x(N/2)	
1	f(1,0)=x(1)+x[(N/2)+1]	f(1,1)=x(1)-x[(N/2)+1]	
*	*	*	(b)
*	*	*	
*	*	*	
(N/2)-1	f[(N/2)-1,0]=x[(N/2)-1]+x(N-1)	f[(N/2)-1,1]=x[(N/2)-1]-x(N-1)	

g(n2,k1)

n2	k1=0	k1=1	
0	f(0,0)	f(0,1)*W_N^0	
1	f(1,0)	f(1,1)*W_N^1	
*	*	*	(c)
*	*	*	
*	*	*	
(N/2)-1	f[(N/2)-1,0]	f[(N/2)-1,1]*$W_N^{(N/2)-1}$	

X(k1+2*k2)

k2	k1=0	k1=1	
0	x(0)	x(1)	
1			
*	*	*	(d)
*	*	*	
*	*	*	
(N/2)-1	X(N-2)	X(N-1)	

Figure 4.7: Computational stages for Eq. (4.23) with N1=2, N2=N/2

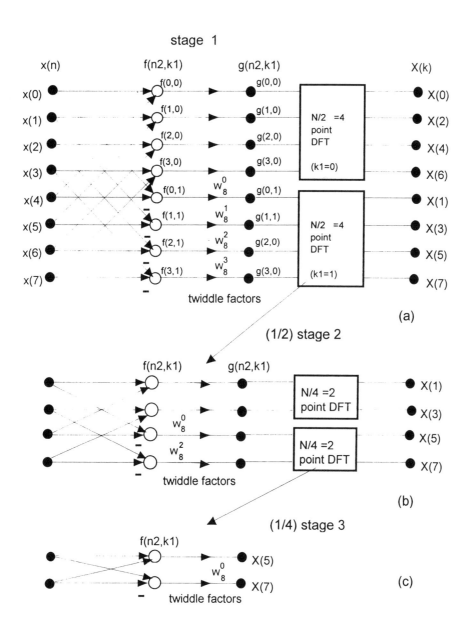

Figure 4.8: Decomposition flow graph for N=8 (2^3)
(a) 8-point DFT into 4-point DFTs
(b) 4-point DFT into 2-point DFTs
(c) 2-point DFT

Since we have carried out the computation in-place, the DFT values are in the order specified by the input index map. Therefore, if we extract the DFT values in the same order that the input was entered into the array (column-wise) then we obtain the even-indexed values first and then the odd-indexed values. For an illustration we take N = 8 which limits n_2 and k_2 indices to the range 0, 1, 2, 3. The corresponding $f(n_2, k_1)$ and $g(n_2, k_1)$ produce the flow graph shown in Fig. 4.8(a) which shows that 8-point DFT is reduced to two 4-point DFT calculations. This is the basic idea behind the radix-2 FFT. The values for $g(n_2, k_1)$ are now stored in the original input data locations because of in-place calculations. Now we have a new set of data divided into two halves to be used as inputs for two 4-point DFTs. The variable n_2 has now the values 0, 1. The corresponding flow graph, for one of the 4-point DFTs, is shown in Fig. 4.8(b). The 4-point DFT is now reduced to a two 2-point DFTs. The new values for $g(n_2, k_1)$ are stored back in the input data locations. The input data are now in four groups each with two values. So, for this case of N = 8, calculations are reduced to four 2-point DFTs. Each of these 2-point DFTs can be examined using again Fig. 4.7 in which n_2 has now only a single value (= 0), but in all cases $0 \le n_1$, $k_1 \le 1$. The final flow graph is shown in Fig. 4.8(c).

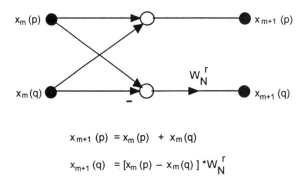

$$x_{m+1}(p) = x_m(p) + x_m(q)$$

$$x_{m+1}(q) = [x_m(p) - x_m(q)] * W_N^r$$

Figure 4.9: Flow graph of a butterfly computation in Fig 4.8

The basic computation at every stage is performed by the flow graph shown in Fig. 4.9, called a butterfly. In general, each butterfly involves one complex multiplication and two complex additions. For N = 2^M, there are (N/2) butterflies per stage (in Fig. 4.8, there are four per stage), and there are M = $\log_2$ N stages. For the FFT, the total number of complex multiplications is (N/2) $\log_2$ N therefore, and the number of complex additions is N $\log_2$ N. (Note that an N-point DFT requires N^2 complex multiplications and N^2 - N complex additions).

In this case we have chosen $N_1 = 2$ and $N_2 = N/2$. The resulting flow graph, Fig. 4.8, is referred to as decimation-in-frequency because the decomposition of the DFT led to smaller and smaller output subsequences. If we had chosen $N_1 = N/2$ and $N_2 = 2$, we would have obtained the decimation-in-time decomposition, it forms smaller and smaller subsequences of the input sequence. There are also other combinations leading to other structures [5,24].

4.5 THE GOERTZEL ALGORITHM

As seen in previous section, the radix-2 FFT algorithm computes the DFT in $(N/2) \log_2 N$ multiplications and $N \log_2 N$ additions for an N-point sequence. If only a selected number of values of the DFT are desired then a direct computation of the desired values is more efficient. Such a direct computation of the DFT is provided by the Goertzel algorithm. It is derived from the DFT, Eq. (4.15), by multiplying it by the factor $W^{-kN} = 1$, and forming

$$X(k) = \sum_{n=0}^{N-1} x(n)\, W^{-k(N-n)} \; ,$$

where W is used, instead of W_N, in order to simplify the notation. This is the form of a convolution, so we can define the sequence

$$y_k(m) = \sum_{n=0}^{N-1} x(n)\, W^{-k(m-n)} \; ,$$

which is the output of a linear network (filter) with the impulse response

$$h_k(n) = W^{-kn}, \quad n \ge 0 \qquad . \tag{4.27}$$

By comparing X(k) with $y_k(m)$ we find that $X(k) = y_k(m=N)$, which means the output of this filter at m=N gives the value of the DFT at the frequency $\omega_k = 2\pi k/N$.

The system or transfer function of the filter defined by Eq. (4.27) is

$$H_k(z) = \frac{1}{1 - W^{-k}\, z^{-1}} \qquad , \tag{4.28}$$

which follows from Table 1.1, for the case c^n with $c = W^{-k}$. The above transfer function has a complex pole $p = W^{-k}$ on the unit circle at the frequency $\omega_k = 2\pi k/N$. Multiplying both the numerator and denominator by $(1 - W^k z^{-1})$, the term for the complex conjugate pole W^k, we obtain

$$H_k = \frac{Y(z)}{X(z)} = \left(\frac{1}{1 - [2\cos(2\pi k/N)]z^{-1} + z^{-2}} \right) \cdot (1 - W^k z^{-1}) \; , \tag{4.29}$$

and so avoid complex multiplications and additions. To obtain the Goertzel algorithm, we partition the above function as follows

$$H_k(z) = \left(\frac{S(z)}{X(z)}\right) \cdot \left(\frac{Y(z)}{S(z)}\right) \quad .$$

(4.30)

Associating now the first and second factors in Eq. (4.30) with the same ones in Eq. (4.29), and converting to the time domain we obtain the Goertzel algorithm as

$$s_k(n) = x(n) + [2 \cos(2\pi k/N)] \, s_k(n-1) - s_k(n-2)$$
$$y_k(n) = s_k(n) - W^k \, s_k(n-1) \quad ,$$

(4.31)

with the initial conditions $s_k(-1) = s_k(-2) = 0$. The desired output is $X(k) = y_k(N)$ for $k = 0, 1, ..., N-1$. The first equation is iterated for $n = 0, 1, ..., N$, but the second equation is computed only once at time $n=N$.

The Goertzel algorithm has advantages for DFT computations at a relatively small number of points, less than $\log_2 N$; otherwise the FFT algorithm is more efficient.

4.6 THE CHIRP Z-TRANSFORM (CZT)

The DFT, Eq. (4.7), can be expressed in terms of the z-transform as follows

$$X(z_k) = \sum_{n=0}^{N-1} x(n) \, z_k^{-n} \quad ,$$

(4.32)

where $z_k = e^{j(2\pi/N)k}$. This shows that the DFT is evaluated at equally spaced points around the unit circle in the z-plane as indicated in Fig. 4.10.

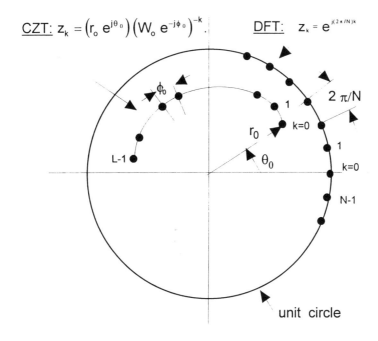

$$\underline{\text{CZT:}} \ z_k = \left(r_0 \, e^{j\theta_0}\right)\left(W_0 \, e^{-j\phi_0}\right)^{-k} . \qquad \underline{\text{DFT:}} \ z_k = e^{j(2\pi/N)k}$$

Figure 4.10: Z-plane contours for the DFT and CZT

The CZT is evaluated along an arbitrary contour, in the z-plane, given by

$$z_k = \left(r_0 \, e^{j\theta_0}\right)\left(W_0 \, e^{-j\phi_0}\right)^{-k} , \qquad (4.33)$$

also shown in Fig. 4.10. The starting point, for k=0, is $r_0 \, e^{j\theta_0}$ and it moves along a contour with increasing value of k towards the origin (if $W_0 > 1$), or away from the origin (if $W_0 < 1$). If $r_0 = W_0 = 1$, the contour is an arc or the unit circle between θ_0 and $(L-1) \, \phi_0$. This arc becomes full circle if $\theta_0 = 0$, $\phi_0 = 2\pi/N$, and L=N; in this case the CZT is the same as the DFT.

Using (4.33) in (4.32) we obtain the CZT as

$$X(z_k) = \sum_{n=0}^{N-1} x(n)\left(r_0 \, e^{j\theta_0}\right)^{-n} W^{kn}, \quad k = 0,1,\dots,L-1 , \qquad (4.34)$$

where $W = W_0 \, e^{-j\phi_0}$. This expression can be arranged in a form of a convolution by using the identity

$$kn = \frac{1}{2}\left[n^2 + k^2 - (k-n)^2\right] \qquad (4.35)$$

Substituting (4.35) into (4.34) we have

$$X(z_k) = W^{k^2/2} \sum_{n=0}^{N-1} \left[x(n)\left(r_0 \, e^{j\theta_0}\right)^{-n} W^{n^2/2} \right] W^{-(k-n)^2/2} \quad . \tag{4.36}$$

Introducing $g(n) = x(n)\left(r_0 e^{j\theta_0}\right)^{-n} W^{n^2/2}$, the above becomes

$$X(z_k) = W^{k^2/2} \sum_{n=0}^{N-1} g(n) \, W^{-(k-n)^2/2} \, , k = 0,1,...,L-1 \quad . \tag{4.37}$$

The summation in (4.37) can be interpreted as the convolution of the sequence g(n) with the impulse response h(n) of a filter given by

$$h(n) = W^{-n^2/2} \quad . \tag{4.38}$$

Therefore, (4.37) can be expressed as

$$X(z_k) = W^{k^2/2} \, y(k) \, , k = 0,1,...,L-1 \quad , \tag{4.39}$$

where y(k) is the output of the filter

$$y(k) = \sum_{n=0}^{N-1} g(n)h(k-n) \quad . \tag{4.40}$$

The sequence $h(n) = W_0^{-n^2/2} \, e^{jn^2\phi_0/2}$, with $W_0 = 1$, is a complex exponential with argument $n^2\phi_0/2 = n(n\phi_0/2)$, where $n\phi_0/2$ corresponds to the frequency $\omega = n\phi_0/2$, increasing linearly with time. This kind of signal, used in radar systems, is known as a chirp signal. This is the reason for naming the z-transform in (4.37) the chirp z-transform.

The linear computation of (4.40) can be done efficiently by use of the FFT, but this has to be considered with care [23, 24, 27].

Both the CZT and Goertzel algorithm perform Fourier analysis as linear filtering. They also enable this analysis to be done over a selected frequency range. However, in addition, the CZT can sharpen resonances by evaluating the z-transform over points away from the unit circle [23].

4.7 SPECTRAL ANALYSIS WITH COMPUTERS

For the frequency analysis using the DFT we have to limit a long signal sequence, x(n), to a finite set of points L. Such a finite observation can be described as passing data through a rectangular window, w(n), which is modelled by the following multiplication

$$\hat{x}(n) = x(n) \cdot w(n)$$
$$w(n) = 1, \qquad\qquad 0 \le n \le \text{L-1} \qquad\qquad (4.41)$$
$$w(n) = 0, \text{otherwise} \quad .$$

Applying the Fourier transform to (4.41) we obtain the convolution in the frequency domain

$$\hat{X}(k) = X(k) * W(k) \quad , \qquad\qquad\qquad (4.42)$$

where X(k) and W(k) are respectively data and window DFT spectra. For a wide window w(n), meaning a large number of data points L, W(k) will be narrow and will have a small effect so that $\hat{X}(k) \simeq X(k)$. But for a narrow window, and correspondingly a small L, W(k) will be wide and will cause frequency spread and distortion, so that $\hat{X}(k) \ne X(k)$.

For the FFT radix-2 algorithm the number of frequency points, N, must be a power of two. If the number of data points L < N, FFT programs add (N-L) zero data points (zero padding).

4.7.1 FFT and Goertzel algorithm tests

To study the FFT operation, data length L, and additive noise effects we have used Matlab and formulated the following program.

```
Matlab program 1 : FFT experiment
L = 256; N = 256;
n = 0 : 1 : L - 1;
x = 0.25 + sin (0.125*pi*n) + 0.3 *sin (0.171875*pi*n) + ...
       0.7 *sin (0.3125*pi*n);
rand ('normal')   % Omit in version 4
y = x + P · randn(1,N);
Y = fft (y, N)
k = 0 : 1 : N - 1;
stem (k, abs(Y))
grid; pause; clg
Y(1), Y(17), Y(23), Y(41)
pause
```

The data signal x(n) of length L consists of dc component and three sinewaves at frequencies 0.125π, 0.171875π, 0.3125π. These frequencies have been chosen to be the 16th, 22nd and 40th harmonics of the FFT frequency interval $2\pi/N = 2\pi/256$ in this case. The finite observation interval for the signal, L, sets a limit on the frequency resolution, i.e. it limits us in distinguishing between two frequency components separated by less than $2\pi/L$ in frequency. This figure comes from the rectangular window as defined in (4.41), whose spectrum W(k) has its first zero crossing at $2\pi/L$ [27, Chapter 6].

In the program shown one can choose the data length L. Choosing L = N = 256, we make the signal appear to the FFT algorithm as periodic and infinitely long (L = ∞), so that the frequency resolution will be perfect. This is seen in Fig. 4.11 showing only the four signal lines. However, decreasing the observation length to L = 200 points we obtain the result in Fig. 4.12 which shows the window effect or so-called spectral leakage.   We see that the signal component at 0.171875π could be lost, especially if its amplitude were smaller than 0.3. There are many types of window which can improve the frequency resolution [6, 12, 27]. The results in Figs. 4.11 and 4.12 are for no noise (P=0). If we include noise then even for L = N = 256, and P = 0.9 we obtain many lines and the 0.171875π component is lost.

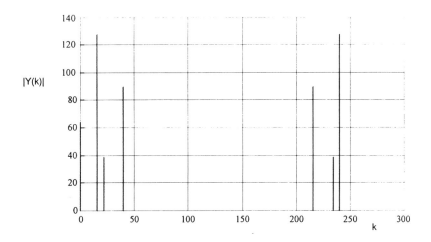

Figure 4.11: Frequency spectrum for L=N=256, P=0 (no noise)

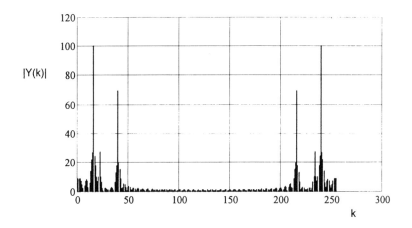

Figure 4.12: As Fig 4.11, but L=200, P=0

For the case L = N = 256, the program 1 gave the FFT values Y(1) = 64, Y(17) = 0.0 - j128, Y(23) = 0.0 - j38.40, Y(41) = 0.0 - j89.60. Note that Matlab indexing starts with 1, so these components are for FFT indexing at 0, 16, 22 and 40.

For the next experiment we use the same signal as in program 1, but instead of the FFT algorithm we use the Goertzel algorithm with the following program as a continuation of program 1.

Matlab program 2 : Goertzel experiment

```
for k = [0   16   22   40 ] ; j = sqrt (-1);
b = [ 1       -exp (-j* 2*pi* k/N)      0] ;
a = [1      - 2 cos (2*pi* k/N)        1] ;
y = filter (b, a, x) ;
y(N),  Y = abs(y (N))
end
```

The coefficients b and a are obtained from Eq. (1.25) for N = 2, and applied to Eq. (4.29). Results for y (N=256) are complex and given as Y(0) = 64, Y(16) = - 48.9 - j118.3, Y(22) = - 19.74 - j32.93, Y(40) = - 74.5 -j49.8, and as such do not agree with the FFT results given earlier. However, the absolute values agree exactly with the FFT results for this case (L = N = 256). The above Goertzel program does not work for L < N; Matlab comment : Index exceeds matrix dimensions.

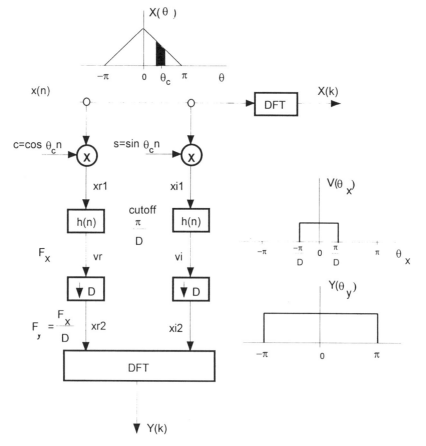

Figure 4.13: Block diagram of the zoom transform

4.7.2 High resolution narrowband spectral analysis

As discussed earlier in this section, the spectral resolution is limited by the data window length L. To resolve close frequency components one would require a long data sequence and a long processing time. An alternative is to use the so-called **zoom transform** as indicated in Fig. 4.13. The signal spectrum $X(\theta)$ is in the range $-\pi \leq \theta \leq \pi$, but we may be interested only in a narrow band of frequencies centred around θ_0. For the zoom transform we first multiply the input signal sequence by $e^{j\omega_c n} = \cos\omega_c n + j\sin\omega_c n$. Then we down sample it by means of the decimator (D) preceded by the lowpass filter with cutoff frequency π/D. (In Section 1.6.1 this cutoff is given as $F/2D$ where F is the sampling frequency of $x(n)$, denoted here as F_x. Then for $f = F_x/2D$, in Fig. 4.13, we have $\theta = 2\pi F_x T_x/2D = \pi/D$, since $F_x T_x = 1$).

The spectrum of the down sampled sequence $y(n)$ is then computed using the DFT. Effectively, down sampling allows us to see the narrowband spectrum on an expanded scale [6, p. 121; 26, p.86].

The method outlined in Fig. 4.13 has been tested on a signal consisting of three closely spaced sinusoidal components at frequencies 0.495π, 0.5π, and 0.505π. The Matlab program used for this experiment is shown below.

```
Matlab program 3 : Zoom transform
n = 1 : 1 : 512 ;
x = sin (0.495*pi* n) + sin (0.5*pi* n) + sin (0.505*pi* n) ;
X = fft (x, 512) ;
k = 1 : 1 : 512 ;
stem (k, abs (X))
grid ; pause ; clg
c = cos (0.5*pi* n) ; xr1 = x. * c ;
s = sin (0.5*pi* n) ; xi1 = x. * s ;
xr2 = decimate (xr1, 10, 'fir');
xi2 = decimate (xi1, 10, 'fir') ;
y = xr2 + xi2 * i          % i = sqrt (- 1)
Y = fft (y, 256)
Y1 = fft shift (Y)
q = - 128 : 1 : 127 ;
stem (q, abs (Y1)).
grid ; pause ; clf
```

Note : In the decimate instruction, the decimation factor D has been chosen to be 10. The associated filter 'fir' within this instruction is an 8-th order Chebyshev filter with the cutoff $0.8\pi/D$, but Matlab uses frequency notation hence the cutoff is 0.8 (fs/2)/D where fs is the sampling frequency, denoted in Fig. 4.13 by F_x.

The input signal spectrum, stem (k, abs(X)), is shown in Fig. 4.14, and the output spectrum stem (q, abs (Y1)) in Fig. 4.15. It demonstrates that the three closely spaced frequency components are now clearly visible. In Fig. 4.15 we have used 256 points, but it has been tried with 64 points, which is just above 512/10 points after decimation, and result was also quite clear, but for 256 points the result is more impressive.

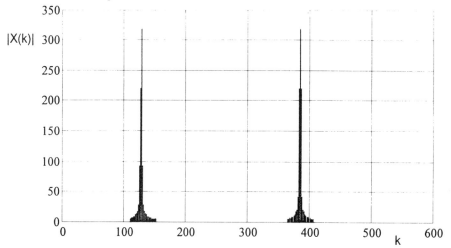

Figure 4.14: Spectrum of a signal consisting of three closely spaced sinusoids

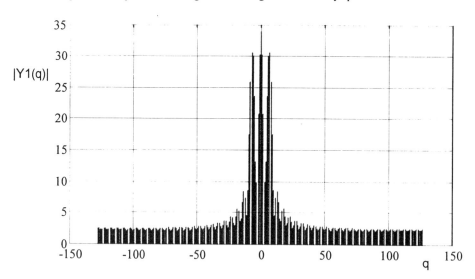

Figure 4.15: Result of the zoom transform applied to fig 4.14

5

DSP implementations of Fourier and Goertzel algorithms

5.0 INTRODUCTION

In this chapter, we shall produce an algorithm in C for calculating an in-place radix-2 DFT, using programming constructions that can be related to the equations from Chapter 4. In Section 5.1 we shall build an intuitive model of the complete algorithm. In 5.2 we will construct the basic algorithm in C, convert it to a form suitable for implementation on **any DSP** and then use it to develop an assembler language program for the TMS320C25 DSP (also compatible with the TMS320C50). In Section 5.3 we show results obtained with the FFT program running on a TMS320 simulator. In 5.4 we follow the same approach to writing an efficient Goertzel algorithm for the DSP and finally Section 5.5 shows some figures computed by both FFT and Goertzel, C and TMS320C25.

5.1 AN INTUITIVE VIEW OF THE FFT

Before attempting to program the FFT, it is helpful to have a physical model of the process. Let us suppose that we have two analogue signals (a) DC and (b) 1 cycle of a sine wave. Two samples A and B are taken of each, as shown in Fig. 5.1. The amplitudes of the DC samples (averaged over the sample time) are both V_{f0}; the sine wave has sample $A = +V_{f1}$ and sample $B = -V_{f1}$. If we compute a two-point FFT using the usual butterfly, we find that the two DC samples give a sum of $2V_{f0}$ and a difference of 0, whereas the sine wave gives a sum of 0 and a difference of $2V_{f1}$ as shown in Fig. 5.2. Thus, in any mixed signal, the butterfly separates out these two components. The subtraction inverts one half cycle of the f1 waveform, effectively rectifying it. Note that the system still works if the sine wave is phase shifted in the time slot.

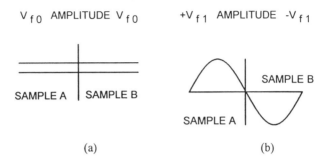

(a)   (b)

Fig. 5.1: (a) Two samples of DC signal (b) Two samples of sine wave

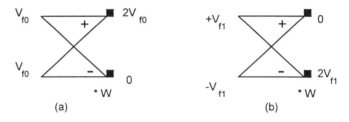

(a)   (b)

Fig. 5.2: Butterfly on (a) DC and (b) Sine waveform

However as phase information is not to be lost, we multiply the result of the subtraction by a twiddle factor which is complex; of course the addition is DC and has no phase information. If the butterfly operates on higher harmonics, even harmonics are enhanced by the addition like a DC voltage and cancelled by the subtraction (Fig. 5.3a). Odd harmonics are enhanced by the subtraction and cancelled by the addition (Fig. 5.3b).

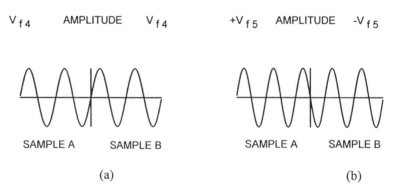

(a)   (b)

Fig. 5.3: (a) Two samples of even harmonic (b) Two samples of odd harmonic

We can use this model of the butterfly to visualise the operations carried out in a 2^M point FFT, and in particular to explain the need for bit-reversed reordering of the output coefficients.

5.2 THE FFT ALGORITHM

Assume that data have been read into a single linear array x[ ] . A general purpose N (= 2^M) point algorithm will be developed. The algorithm to compute an eight point FFT (M=3) is shown in Fig. 5.4. (compare with Fig. 4.8). Each one of the transforms in the three stages corresponds to the operations described in Fig. 4.7 where $x(n) = x(\frac{N}{2}n_1 + n_2)$ is transformed to $X(k_1 + 2k_2)$. The even set of $X(k_1 + 2k_2)$ and the odd set correspond to the separation of even and odd harmonics described in Section 5.1. In our algorithm, the first stage converts 8 samples to 2 sets of 4 and finally, in the third stage to 4 sets of two. Taking the first stage (m=0), four butterflies are computed. Imagine that x[0] to x[3] hold sample A from Fig. 5.1 and x[4] to x[7] sample B. Butterfly additions are stored in x[0] to x[3], so that only the even harmonics can be here including the DC component, which is distributed between x[0] to x[3]. The DC component is extracted in the m=1 stage by repeated addition, stored in x[0], x[1] which are finally added in the last stage and stored in x[0]. Note that all AC components have been cancelled by this stage. Butterfly subtractions in the first (m=0) stage are stored in x[4] to x[7], so that the odd harmonics are here, including the fundamental f1, 'rectified' and distributed between x[4] to x[7]. One might think of f1 as the DC component of the odd harmonic samples x[4] to x[7] at the second (m=1) stage and indeed, additions at this stage store the f1 component in x[4], x[5], which are added in the last stage to give the f1 coefficient in x[4]. One can envisage all the component frequencies being extracted in this way. The binary output index in Fig. 5.4 shows the building up of the output coefficient index in bit-reversed order. After the first stage, x[0] to x[3] hold even harmonics (bit 0 = 0) and x[4] to x[7] the odd ones (bit 0 = 1). In the second (m=1) stage the odd and even harmonics in each half are again separated, giving bit 1 of the binary output coefficient index. It can now be seen where the final coefficients will be stored at the last stage and why the location can be found by bit-reversing the original data sample indices.

Ntd2 = N/2 = 4 for(n2=0;n2<Ntd2;n2=n2+1) n2=		Nt = N = 8 for(i=n2;i<N;i=i+Nt) i =	x[i], x[i+Ntd2]	BUTTERFLY SEQUENCE	X[k] k =	BINARY OUTPUT INDEX
0	x[0]	■ 0	+	1	0	xx0
1	x[1]	■ 1	+	2	2	xx0
2	x[2]	■ 2	+	3	4	xx0
3	x[3]	■ 3	+	4	6	xx0
m=0 0	x[4]	■ 4	- * W0	1	1	xx1
1	x[5]	■ 5	- * W1	2	3	xx1
2	x[6]	■ 6	- * W2	3	5	xx1
3	x[7]	■ 7	- * W3	4	7	xx1

Ntd2 = N/4 = 2 for(n2=0;n2<Ntd2;n2=n2+1) n2=		Nt = N/2 = 4 for(i=n2;i<N;i=i+Nt) i =	x[i], x[i+Ntd2]	BUTTERFLY SEQUENCE	X[k] k =	BINARY OUTPUT INDEX
0	x[0]	■ 0	+	5	0	x00
1	x[1]	■ 1	+	7	2	x00
0	x[2]	■ 0	- * W0	5	1	x10
1	x[3]	■ 1	- * W2	7	3	x10
m=1 0	x[4]	■ 4	+	6	0	x01
1	x[5]	■ 5	+	8	2	x01
0	x[6]	■ 4	- * W0	6	1	x11
1	x[7]	■ 5	- * W2	8	3	x11

Ntd2 = N/8 = 1 for(n2=0;n2<Ntd2;n2=n2+1) n2=		Nt = N/4 = 2 for(i=n2;i<N;i=i+Nt) i =	x[i], x[i+Ntd2]	BUTTERFLY SEQUENCE	X[k] k =	BINARY OUTPUT INDEX
0	x[0]	■ 0	+	9	0	000
0	x[1]	■ 0	- * W0	9	1	100
0	x[2]	■ 2	+	10	0	010
0	x[3]	■ 2	- * W0	10	1	110
m=2 0	x[4]	■ 4	+	11	0	001
0	x[5]	■ 4	- * W0	11	1	101
0	x[6]	■ 6	+	12	0	011
0	x[7]	■ 6	- * W0	12	1	111

Fig. 5.4: In-place FFT, N=8 showing data flow in program

5.2.1 The iterative structure of the FFT program

We now need to develop a program to perform the FFT of Fig. 5.4; we start with the iterative structure. We shall build a program to compute an N-point FFT where $N = 2^M$. An iterative scheme is used, so that N can easily be changed. The steps in the FFT algorithm are:

- Read data into array
- Perform DFT
- Reorder coefficients (bit reversal)
- Output data

On a DSP, all these steps apart from the transform are quite straightforward and we shall concentrate on a DFT algorithm written in C.

```
for(m = 0;m < M; m = m+1){ /* M sets of FFT */
    for(n2 = 0;n2 < Ntd2;n2 = n2 + 1){
        /* loop through FFT points */
        /* get new twiddle factor */
        for(i = n2;i < N;i = i + Nt){ /* loop through FFTs */
            /* compute butterfly on x[i], x[i+Ntd2] */
        }
    }
}
```

Fig. 5.5: Iteration for an N $(=2^M)$ point in-place FFT algorithm

The three level iteration to perform the transform of Fig. 5.4 in C is shown in Fig. 5.5. The outer loop starts with an Nt $(=8)$ point transform and steps through M stages (3 in Fig. 5.4) with each successive stage having twice the number of transforms each of half the size (Nt=8,4,2). Nt corresponds to N in Fig. 4.7(a) and Ntd2 = Nt/2. The butterfly therefore starts with samples x[i], x[i+Ntd2], corresponding to x(0),x(N/2) in Fig. 4.7a. The resulting values of X(0), X(1) are stored in x[i], x[i+Ntd2]. The two inner iterations step through all the points of each FFT. However, instead of completing each transform in turn, the first point of every one is calculated (inner loop: i = n2;i < N; i=i+Nt) before going on to the second point (outer loop: n2=0;n2<Ntd2;n2=n2+1). The progress of indices n2 and i as each butterfly is computed can be followed in Fig. 5.4. Thus

the butterfly sequence shows that in the m=0 stage the first operations are on x[0], x[4], followed by x[1], x[5] etc. The reason for this 'unnatural' sequence can be seen by looking at the sequence of twiddle factors W0 to W3. This sequence minimises the evaluations of the twiddle factors, e.g. at each stage, every operation involving W0 is completed before calculating W1. The binary output index shows the location of the coefficients after each stage, as described in the previous section. By the Mth stage the coefficients are in bit-reversed order.

5.2.2 The butterfly

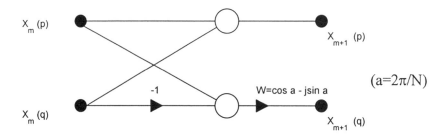

Fig. 5.6: In-place butterfly

The same piece of C code is used for every butterfly calculation, as shown in Fig. 5.6. The input and output data are complex and each real-imaginary pair is stored in adjacent elements of array x[ ]. Thus each element has a size (ES) of 2. The iterative sequences of Figs. 5.4 and 5.5 were, for clarity demonstrated with ES = 1. Converting the butterfly of Fig. 5.6 into a C-like form, we have

$X_{m+1}(p) = X_m(p) + X_m(q)$ ≡ $\{x[i] + x[l]\}_m + j\{x[i+1] + x[l+1]\}_m$

→ $\{x[i]\}_{m+1} + j\{x[i+1]\}_{m+1}$

$X_{m+1}(q) = (X_m(p) - X_m(q))W$ ≡ $(\{x[i] + j x[i+1]\}_m - \{x[l] + j x[l+1]\}_m) * (\{cos(a) - j \; sin(a)\}_m)$

≡ $(\{x[i] - x[l]\}_m + j\{x[i+1] - x[l+1]\}_m) * (\{cos(a)\}_m - j\{sin(a)\}_m)$

→ $(\quad tr \quad + j \quad ti \quad) * (\{cos(a)\}_m - j\{sin(a)\}_m)$

≡ $tr * \{cos(a)\}_m + ti * \{sin(a)\}_m + j (ti * \{cos(a)\}_m - tr * \{sin(a)\}_m$

→ $x[l]_{m+1} \quad\quad + j \quad\quad x[l+1]_{m+1}$

```
/* e = angle between cos/sin table entries */
#define ES 2 /* element size */
for(m = M-1;m < M; m = m +1){ /* M sets of FFT */
    a = 0.0;
    for(n2 = 0;n2 < Ntd2;n2 = n2 + ES){ /* points */
        sina = sin(a);cosa = cos(a); /* DSP uses table */
        a = a + e;
        for(i = n2;i < ES * N;i = i + Nt){ /* IFFTs */
            l = i + Ntd2;
            tr = x[i] - x[l];    /* butterfly */
            x[i] = x[i] + x[l];
            ti = x[i+1] - x[l+1];
            x[i+1] = x[i+1] + x[l+1];
            x[l+1] = cosa * ti - sina * tr;
            x[l] = cosa * tr + sina * ti;
        }
    }
    Nt = Ntd2; Ntd2 = Ntd2/ 2; /* split lengths */
    e = 2 * e; /* double twiddle factor angle */
}
```

Fig. 5.7 (a): In-place N-point FFT program in C using an indexed array

Three changes are needed to convert the C code into a form suitable for evaluation on a DSP. Firstly the data are complex and read into x[ ] in the order real, imaginary, real, imaginary etc. Thus each element of the array has size 2 and we have defined a constant ES = 2. Secondly, assembler languages give efficient support to an iterative structure equivalent to do{....}while, where the modify and test of the loop variable is at the end. Thirdly, we must convert the butterfly to a form using indirect addressing where the efficient post-increment DSP addressing modes can be employed.

The indexed array version shown in Fig 5.7a is converted to the pointer addressed one as shown in Fig. 5.7b. The pointers pi and pl are used to step through x[ ]. The inner loop is now terminated when pointer pi exceeds the address in the array where the last data are stored. This algorithm has been executed with good results for values of N between 64 and 512.

```
/* e = angle between cos/sin table entries */
 m = M-1;
do{ /* M sets of FFT */
    a = 0.0;
    n2 = 0;
    do{ /* loop through FFT points */
        sina = sin(a);cosa = cos(a) /* DSP uses table */
        a = a + e;
        pi = &x[n2]; /* set low pointer */
        pl = &x[n2 + Ntd2]; /* set high pointer */
        do{ /* loop through FFTs */

            tr = *pi - *pl;    /* butterfly */
            *pi++ = *pi + *pl++;
            ti = *pi - *pl;
            *pi-- = *pi + *pl;
            *pl-- = cosa * ti - sina * tr;
            *pl = cosa * tr + sina * ti;
            pi = pi + Nt;
            pl = pl + Nt;
        }while(pi < (&x[0] + ES * N));
        n2 = n2 + ES;
    }while(n2 < Ntd2); /* end of n2 loop */
    Nt = Ntd2; Ntd2 = Ntd2/ 2; /* split lengths */

    e = 2 * e; /* double twiddle factor angle */
    m = m - 1;
}while(m >= 0); /* end of m loop */
```

Fig. 5.7 (b): In-place N-point FFT program in C using pointer addressing

5.2.3 The assembler language FFT

A brief description of the assembler directives, addressing modes and instruction subset of the TMS320C25 is given in Appendix 5.1. Refer to [4,29] for more details. A TMS320C25 program 'FFT256' derived from the C prototype is given in full in Appendix 5.2 and performs a 256 point transform. The auxiliary registers AR6 and AR5 correspond to pointers pl and pi respectively. The C statements of Fig. 5.7b are used as comments in the assembler version where the corresponding functions are performed. Cosine and sine functions for the twiddle factors are easily performed in look-up tables in assembler languages. The reordering of the coefficients is easily performed with the reverse carry addressing mode of the DSP. Performing a square root operation on the complex coefficients to obtain magnitudes is a time-consuming operation which can usually be avoided in tasks using FFTs computed in real time. Therefore the complex coefficients and the power spectrum (real^2 + imaginary^2) are output.

FFT256 is not particularly optimised for speed. For example, the bit reversal and output section could be rewritten to run at least one machine cycle faster per iteration, with some loss of program structure. Constants used in 'immediate' operations could be stored in data memory to gain speed. By altering N and M, the number of points can be changed, without major program modification.

Butterfly multiplications cannot overflow, if the extra sign bit in the product of the twos complement fractional multiplication has been shifted out (see Section 2.2.3), in accordance with normal practice. The additions and subtractions in the butterfly can give rise to overflow, however. The + and - operations in the FFT butterfly are performed thus on the TMS320:

```
lac *,shift15,ar5
add *,shift15,ar6
sach *-,scale,ar5    ;*pi-- = *pi + *pl
```

The symbol 'scale' is a left shift value and would be 1 to give no scaling, as the arithmetic is carried out in bits 15 to 30 of the accumulator. In a 64 point transform, without scaling, a peak sine input of 1000 ($\equiv$ 0.03 or 1000/32768) is the maximum that the program can handle without overflow occurring. In Appendix 5.2, scale has been set to zero, equivalent to a division by 2 at each butterfly. This has improved the dynamic range so that the full input of $\pm$ 30000 ($\equiv$ $\pm$ 1) can be handled. The reason for the apparently strange tactic of not performing the add and subtract operations in either bits 0-15, or bits 16-31 of the accumulator can now be seen.

5.3 SOME EXPERIMENTS WITH THE DSP FFT PROGRAMS

Results may be obtained conveniently from DSP-based algorithms by execution on a simulator. A TMS320C25 simulator [4], running under MSDOS was used to obtain the results which follow. A major advantage compared to the real silicon is that input data may be read from computer generated files of precise and repeatable data. Output is also easily captured on disc and can easily be displayed using standard graphic tools.

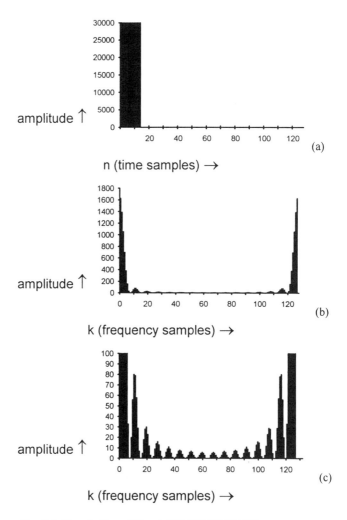

Fig. 5.8: Results from 128 point TMS320C25 FFT program
(a) Input pulse of length L=16
(b) Power spectrum output
(c) Magnified output showing L repeat pattern

Experiment 1:

A 128 point TMS320C25 FFT program similar to that of Appendix 5.2 was used to generate the DFT from a pulse of length 16; the input data are shown in Fig. 5.8a. For a finite sequence of length L, and unity amplitude, the magnitude of an N-point DFT is given by

$$\left|X(k)\right| = \frac{\sin(\pi kL/N)}{\sin(\pi k/N)} \quad ,$$

where $N > L$ [27, Chapter 6, p442]. Zeros of this function are at $k = m(N/L)$, where $m = 1,2,......$ For Fig. 5.8, with $L = 16$ and $N = 128$, we have $k = 8m$, with $m = 1,2,.......15$.

Experiment 2:

The FFT program of Appendix 5.2 can be extended to 512 points, without major alterations, by using off-chip memory to hold the x array. This TMS320C25/C50 FFT demonstration uses an input file of three sampled waveforms added to obtain simulator input data as follows:

(1) 501 points of sine waveform of peak amplitude ±8000, $L = 501$, $N = 512$
(2) 501 points of sine waveform of peak amplitude ±8000, $L = 501$, $N = 512$
(3) Random numbers uniformly distributed in the range (-16000,+16000)

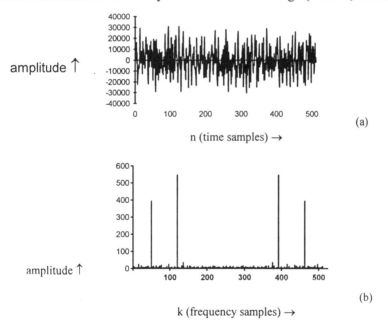

(a)

(b)

Fig. 5.9: Results from 512 point TMS320C25 FFT
    (a) Input signal (two sine waves fs/f=50, amplitude ±8000 and fs/f=120,
        amplitude ±8000 + random noise distributed between -16000 and +16000)
    (b) output

The results are shown in Fig. 5.9 and show an impressive extraction of the sinusiodal components from the noise.

For a finite length L of sine waves of unity amplitude and frequency ω_i, the N-point DFT magnitude can be shown to be

$$|X(k)| = \frac{1}{2} \frac{\sin[\pi(k-k_i)\frac{L}{N}]}{\sin[\pi(k-k_i)\frac{1}{N}]} \;,$$

where the frequency we have used $\omega_i = 2\pi k_i / N$.

For $k = k_i$, the ratio becomes 0/0, and by differentiating the numerator and denominator (L' Hospital rule) and setting $k = k_i$, we obtain $|X(k_i)| = L/2$.

In experiment 2, $\omega_1 = 2\pi\, 50$, $\omega_2 = 2\pi\, 120$, so with $N = 512$ we have integers $k_1 = 25600$ and $k_2 = 61440$ (obtained from $\omega_i = 2\pi k_i/N \rightarrow k_i = \omega_i N/2\pi$ (i = 1,2). For $N = 512$ and $\omega_i = 2\pi \times 50$, $k_1 = 25600$, $\omega_2 = 2\pi \times 120$, $k_2 = 61140$). Similarly to experiment 1, we can obtain zeros for (b) and express zero points as

$$k = k_i \pm m\, N/L \cong k_i \pm m$$

since $L \approx N$. Therefore, zeros are at $k = k_i \pm m$, m = 1,2,... but for m = 0, there is a peak = L/2 as derived in the above.

5.4 GOERTZEL IMPLEMENTATION ON A COMPUTER

We shall implement the Goertzel algorithm as a conventional C language program at first, as in the case of the FFT. Again we shall then modify the code to exploit the features of DSPs and improve execution speed.

Starting from the Goertzel algorithm, Eq. 4.31, we have $s_k(n)$ iterated from 0 to N for N samples and $y_k(n)$ calculated just after the final iteration. For practical implementation, we will arrange them as follows:

ak = 2cos(2πk/N)   bk = e⁻ʲ(2πk/N)

$$= \cos(2\pi k/N) - j \sin(2\pi k/N)$$

$$= bkr - j\, bki, \quad \text{where } bkr = \cos(2\pi k/N),\; bki = \sin(2\pi k/N)$$

| sk(n) = x(n) + ak sk(n-1) - sk(n-2)   (5.1a)

                    iterate n = 0,1.......N-1   .

Now the last iteration:

| sk(N) = x(N) + ak sk(N-1) - sk(N-2)

| = 0   + ak sk(N-1) - sk(N-2)   (5.1b)

| yk(N) = sk(N) - bk sk(N-1)   (5.2a)

| = sk(N) - bkr sk(N-1) + j bki sk(N-1) .   (5.2b)

As the intention is to implement these equations in C, we have modified the symbol names into more C-like terms. Outside the range of N input samples used by a FFT, x(n)=0. We can show an implementation of this in C on an array of x[N] samples in Fig. 5.10.

```
for(n = 0;n < N;n = n + 1){ /* iterate Eq. 5.1a N times */
        skn = x[n] + ak * sknm1 - sknm2;
        sknm2 = sknm1; sknm1 = skn;  /* update sk values */
}
skn = 0 + ak * sknm1 - sknm2;   /* (N+1)th iteration with x(n)=0 */
*re = skn - bkr * sknm1; /* Eq. 5.2b, get real & imaginary parts */
*im = bki * sknm1;
```

Fig. 5.10: Basic Goertzel algorithm in C

5.4.1 Efficient DSP Goertzel implementation

In order to improve the program of Fig. 5.10 for efficient DSP execution we shall make the following changes:

- To avoid the overhead of copying the sknm1, sknm2 values, we will swap the use of these variables each time that Eq. 5.1a is executed and copy the computed value of sk(n) to the sk(n-2) location. Therefore we have N/2 iterations of odd + even evaluations of Eq. 5.1a.
- Use pointers to step through the array with the efficient DSP post-increment addressing mode.
- Combine the last iteration of Eq. 5.1a with the evaluation of Eq. 5.2b.

Combining equations 5.1b & 5.2b and noting that bkr = ak/2, we find

$$yk(N) = ak/2 \; sk(N-1) - sk(N-2) + j \; bki \; sk(N-1) \quad . \tag{5.3}$$

The essential part of a C implementation of this is shown in Fig. 5.11

```
for(n = N/2;n > 0;n = n -1 ){              /* iterate N times */
    sknm2 = *px++ + ak * sknm1 - sknm2;   /*even:  sknm1=sk(n-1) sknm2=sk(n-2) */
    sknm1 = *px++ + ak * sknm2 - sknm1;   /* odd: sknm2=sk(n-1) sknm1-sk(n-2) */
}
*re = bkr * sknm1 - sknm2;       /* combine last iteration with */
*im = -bki * sknm1;              /* output equation, x(n)=0 */
```

Fig. 5.11: DSP-style Goertzel algorithm

5.4.2 Goertzel in DSP assembler code

The algorithm of Fig. 5.11 converts into TMS320C25 assembler code without any particular difficulty, except for the problem of dynamic range, on which we shall have more to say later. We expect this in such structures with feedback, i.e. IIR filters. A listing is given in Appendix 5.3. Although this program uses an array of values stored in memory as the input, unlike the FFT case this need not be so. One of the advantages of the Goertzel algorithm is that computing can start as soon as the first input sample is read and samples need not be stored. This means that, compared to the FFT, memory to store the whole sequence of input data is not needed and also that, because most of the computing could be performed immediately on incoming data, the spectral coefficients(s) $X(k) = y_k(N)$ could be computed very soon after the last sample just by computing Eq. 5.3.

5.5 GOERTZEL AND FFT RESULTS

Experiment 3:

This is the case also considered in Chapter 4,   Section 4.7.1. The TMS320C25 Goertzel program of Appendix 5.3 was used to extract the coefficients for k=0, k=16, k=22 k=40 from the signal:

$$x(n) = 0.25 + \sin(0.125\ \pi n) + 0.3\ \sin(0.171875\ \pi n) + 0.7\ \sin(0.3125\ \pi n)\quad .$$

The file representing the input waveform must occupy the range ±32767, so that a scaling factor of 32767/2.25 has been applied to give the maximum amplitude within these bounds.

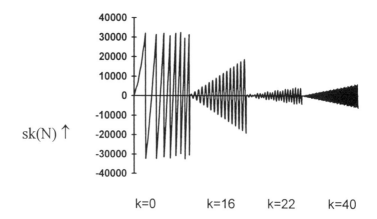

Fig. 5.12: Eq. 5.1a output for k=0, 16, 22 & 40
during 256 iterations of the Goertzel algorithm (TMS320C25)

The critical point for overload of the integer processor is the node where Eq. 5.1a is iterated (skn in Fig. 5.10). Fig. 5.12 shows the values of this node, sk(n), during the computation of the four components of this signal on the TMS320C25.

It can be seen that there is a large range in sk(n), depending on the value of k. In fact, it looks as though when k=0 (DC), there is severe overflow, with large positive numbers turning into large negative ones. This is in fact the case, nevertheless, results are reasonably accurate, as shown in Fig. 5.13.

	k=0	k=16	k=22	k=40
FFT (C) real	931840	-1	2	1
FFT (C) imaginary	0	-1875517	-559154	-1304799
GOERTZEL (C) real	931896	15	3	11
GOERTZEL (C) imaginary	0	-1875517	-559154	-1304800
FFT(DSP)real	3639	0	0	0
FFT(DSP) imaginary	0	-7318	-2177	-5085
GOERTZEL(DSP) real	3491	51	12	20
GOERTZEL(DSP) imaginary	0	-7317	-2192	-5098

Fig. 5.13: Raw output values from FFT and Goertzel programs
in C and TMS320C25 assembler code for k=0,16,22,40

How can this be? The reason is that, in a twos complement number system, overflow in intermediate results of an iterated accumulation have no effect as long as the final result is within range. Obviously high order digits do not affect the final result, so that losing them does not matter. It is important to use this property of an integer arithmetic scheme as dynamic range is always too small in practical integer systems.

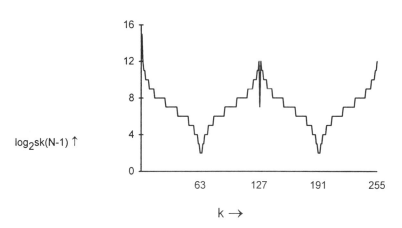

Fig. 5.14: Log sk(N-1) magnitudes for all values of k, (N=256)
for Goertzel algorithm input $x(n) = 1 + \sin(2\pi k \, n/N)$

With this in mind, Fig. 5.14 shows the final value of sk(n) (on a logarithmic scale) at the end of each 0 to N-1 iteration of the first equation of the Goertzel algorithm (N=256).

The input used to produce these results was a DC component of 1 and a sine wave of amplitude 1 appropriate to each value of k. Although the sine wave amplitudes computed were all the same, the intermediate values of sk(n) can be seen in Fig. 5.14 to vary enormously. Most integer systems cannot handle this range.

One useful strategy here is known as block floating point. In such a scheme, different scaling factors are used for each block of computations. In this case, Fig. 5.14 shows that the input can be scaled appropriately according to the k value being calculated. So the TMS320C25 Goertzel program of Appendix 5.3 uses the facility to shift by a **variable** in register T (using the LACT instruction). The subroutine caller can load the T register with a scaling factor appropriate to the desired k value.

A final comment must be made on the speed of execution of the Goertzel versus the FFT algorithm. The TMS320C25 256 point FFT algorithm executes in around 55,000 machine cycles (giving 256 coefficients) compared to about 2200 for Goertzel (giving one coefficient). The Goertzel algorithm may therefore be the fastest option where only a few coefficients are needed. In addition, both the FFT and Goertzel algorithms run into dynamic range problems where a large number of data points are used and computation is performed in integer arithmetic. Under these circumstances, it may be practical to use double precision addition and software floating point multiplication in Goertzel when it would be impractical for the FFT.

APPENDIX 5.1: TMS320C25 ASSEMBLER AND INSTRUCTION SUMMARY

The architecture of the TMS320C25 is summarised in Section 3.3. The TMS320C50 supports all TMS320C25 instructions (source code compatible) although a somewhat modified assembler language is often used. See the TMS320C25 or TMS320C50 User Guides from Texas Instruments for full descriptions of the architectures and instruction sets.

On-chip data memory

For efficient execution of some instructions, on-chip data memory must be used. Examples are all instructions that include the DMOV data move operation, e.g. the MACD convolution operation. The area 300H to 3FFH is normally preferred. The on-chip area 200H to 2FFH can be transformed into read/write program memory at FF00H by the CNFP instruction; this has been used in the adaptive filter listed in Appendix 8.2.

Assembler directives

The TMS320C25 assembler was written at Birmingham University and uses the following directives:

	origin <address>		;set next program memory address
	ramaddr <address>		;set next data memory address (for ramspace)
<label>	**romspace**	<N>	;reserve N 16-bit words in program memory
<label>	**ramspace**	<N>	;reserve N 16-bit words in data memory
<symbol>	**equ**	<N>	;create symbol with 16-bit value N
<label>	**include**	'filename'	;insert named source file
	h.<symbol>		;extract 9 high bits of symbol
			;(to load to data page pointer)
	l.<symbol>		;extract 7 low bits of symbol
			;(to use in direct addressing)

See Section 3.3 for a description of TMS320C25/C50 direct addressing

TMS320C25/C50 instruction set summary

	TMS320C25 indirect addressing modes
*	use the address in the current AR
*+	as above but add 1 to the current AR *after* instruction execution
*-	as above except subtract 1 from the current AR *after* instruction execution
*0+	use current AR but add AR0 to the current AR *after* instruction execution
*0-	as above except subtract AR0
*BR0+	as *0+ except the carry in the AR add is in the reverse direction
*BR0-	as *0- except the carry in the AR sub is in the reverse direction.

The following is a summary of the TMS320C25 instruction set.
A full description can be found in the Texas Instruments User's Guide [29]

Accumulator operations		
MNEMONIC	WORDS/ CYCLES	ACTION sh is a left shift, k8=0-FFH, k16=0-FFFFH, dma is any indirect addressing mode or the low 7 bits of a data memory address (9 high bits come from DP)
ABS	1/1	if ACC < 0 ACC = - ACC
ADD dma,sh	1/1	ACC=ACC+dma contents (sign extended) sh=0-15
ADDC dma	1/1	ACC=ACC+dma contents + carry flag
ADDH dma	1/1	ACC high=ACC high+dma contents (ACC low unaffected)
ADDK k8	1/1	ACC=ACC+k8, k8=0-255
ADDS dma	1/1	ACC=ACC+dma contents (not sign extended)
ADDT dma	1/1	ACC=ACC+dma contents (sign extended) shift in T register
ADLK k16,sh	2/2	ACC=ACC + k16 (sign extended),sh=0-15,k16=0-FFFFH
AND dma	1/1	ACC=ACC and dma contents (no sign extension)
ANDK k16,sh	2/2	ACC=ACC and k16 (*not* sign extended)sh=0-15,
CMPL	1/1	ACC= not ACC
LAC dma,sh	1/1	ACC=dma contents (sign extended) sh=0-15
LACK k8	1/1	ACC=k8 , k8=0-255
LACT	1/1	ACC=dma contents (sign extended) shift in T register
LALK k16,sh	2/2	ACC=k16 (sign extended), sh=0-15, k16=0-FFFFH
NEG	1/1	ACC = -ACC
NORM dma	1/1	if(ACC bit31 = ACC bit 30) ACC = ACC * 2
OR dma	1/1	ACC=ACC or dma contents (no sign extension)
ORK k16,sh	2/2	ACC=ACC or k16 (*not* sign extended), sh=0-15
ROL	1/1	rotate ACC left through carry flag
ROR	1/1	rotate ACC right through carry flag
SACH dma,sh	1/1	store left shifted ACC high to dma (sh=0-7)
SACL dma,sh	1/1	store left shifted ACC low to dma (sh=0-7)
SBLK k16,sh	2/2	ACC=ACC-k16 (sign extended), sh=0-15,k16=0-FFFFH
SFL	1/1	left shift ACC by 1 bit (ACC=ACC× 2)
SFR	1/1	arithmetic shift right by 1 bit (ACC=ACC/2)
SUB dma,sh	1/1	ACC=ACC-dma contents (sign extended) sh=0-15
SUBB dma	1/1	ACC =ACC -dma contents - NOT carry flag
SUBC dma	1/1	use with RPTK for long division
SUBH dma	1/1	ACC high=ACC high-dma contents (ACC low unaffected)
SUBK k8	1/1	ACC=ACC-k8, k8=0-255
SUBS dma	1/1	ACC=ACC-dma contents (not sign extended)
SUBT dma	1/1	ACC=ACC-dma contents (sign extended) shift in T register

TMS320C25/C50 instruction set summary - continued

		Auxiliary register (AR, ARP) and data page pointer (DP) operations
MNEMONIC	WORDS/ CYCLES	ACTION ARn represents AR0, AR1....AR7, dma is any indirect addressing mode or the low 7 bits of a data memory address (9 high bits come from DP)
XOR dma	1/1	ACC=ACC xor dma contents (no sign extension)
XORK k16,sh	2/2	ACC=ACC xor k16 (*not* sign extended)sh=0-15,
ZAC	1/1	LACK 0
ZALH dma	1/1	ACC high= dma contents, ACC low=0
ZALR dma	1/1	ACC low= dma contents with rounding, ACC high=0
ZALS dma	1/1	ACC low= dma contents, ACC high=0
ADRK k8	1/1	ARn = ARn + k8, k8=0-255
CMPR eq	1/1	if Arn eq(als) AR0, TC flag = 1, eq may be lg, gt, ne
LAR Arn, dma	1/1	ARn = dma contents
LARK ARn,v	1/1	ARn = v; v <= FFH
LARP q	1/1	load the auxiliary register pointer with q (0<=q <=7)
LDP dma	1/1	load the auxiliary register pointer from memory
LDPK p	1/1	load the data page pointer with the page no.p (p < 1FFH)
LRLK ARn,v	2/2	ARn = v; v <= FFFFH
MAR dma	1/1	dummy instruction to modify ARP, dma=indirect mode
SAR Arn, dma	1/1	Store ARn in memory
SBRK k8	1/1	ARn - ARn - k8, k8=0-255

		Multiply/accumulate operations
MNEMONIC	WORDS/ CYCLES	ACTION dma is any indirect addressing mode or the low 7 bits of a data memory address (9 high bits come from DP)
APAC	1/1	ACC = ACC + P (with possible shift)
LPH dma	1/1	P reg bits 16-31 = memory contents
LT dma	1/1	T = dma contents
LTA dma	1/1	LT and ACC=ACC+P
LTA dma	1/1	LT and ACC=ACC+P
LTD dma	1/1	LTA and copy dma contents to dma+1
LTP dma	1/1	LT and ACC=P
LTS dma	1/1	LT and ACC=ACC-P
MAC pma, dma	2/(1)	use with RPTK, pma=addr of table in program memory incremented each repeat, dma = indirect e.g. *-. N repeats takes (N+2 cycles). ACC=ACC+P, P=table data× dma data
MACD	2/(1)	as MAC with DMOV
MPYA dma	1/1	ACC = ACC + P (with shift), *then* MPY
MPYK k13	1/1	P = T x k13, k13= 13 bit signed constant
MPYS dma	1/1	ACC = ACC - P (with shift), *then* MPY
MPYU	1/1	unsigned multiply P = T × dma contents
PAC	1/1	ACC = P (with shift)
SPAC	1/1	ACC = ACC - P (with shift)
SPH dma	1/1	Store P reg bits 16-31 to memory
SPH dma	1/1	Store P reg bits 0-15 to memory
SPM s	1/1	set P → ACC shift s (0=0, 1=left1, 2=left2, 3=right 6)
SQRA dma	1/1	ACC = ACC + P, P = dma contents × dma contents (signed multiply)
SQRS dma	1/1	ACC = ACC - P, P = dma contents × dma contents (signed multiply)

TMS320C25/C50 instruction set summary - continued

MNEMONIC	WORDS/ CYCLES	ACTION dma is any indirect addressing mode or the low 7 bits of a data memory address (9 high bits come from DP)
		Memory operations
BLKD dma, dma	2/(1)	use with RPTK, dma=addr of data in data memory incremented each repeat, dma = indirect e.g. *+. N repeats takes (N+2 cycles). copies dma contents to dma
BLKP pma, dma	2/(1)	use with RPTK, pma=addr of data in program memory incremented each repeat, dma = indirect e.g. *+. N repeats takes (N+2 cycles). copies pma contents to dma
DMOV dma	1/1	copy dma contents to dma+1 (in one cycle! on chip only
IN dma	1/2	input port data to data memory
OUT dma	1/2	output data from data memory to port
TBLR dma	1/3	copy program memory (addr in ACC) to dma
TBLW dma	1/3	copy dma (addr in ACC) to program memory

MNEMONIC	WORDS/ CYCLES	ACTION (all branches are absolute to a 16 bit address). All branches may be followed by an indirect addressing mode
		Transfer of control operations
B addr	2/3	unconditional branch to 16 bit address
BANZ addr	2/3	branch if current AR is not 0 decrement AR
BGEZ addr	2/3	branch if ACC >= 0
BGZ addr	2/3	branch if ACC > 0
BLEZ addr	2/3	branch if ACC <= 0
BLZ addr	2/3	branch if ACC < 0
BNV addr	2/3	branch if OV=0
BACC addr	2/3	branch to addr in ACC
BBNZ addr	2/3	branch if TC flag ≠ 0
BBZ addr	2/3	branch if TC flag = 0
BC addr	2/3	branch if carry flag = 1
BNC addr	2/3	branch if carry flag ≠ 1
CALA	2/3	push current PC to stack and branch to address in ACC
BNZ addr	2/3	branch if ACC is not zero (no zero flag)
BV addr	2/3	branch if OV=1 *and reset overflow flag*
BZ addr	2/3	branch if ACC is zero (no zero flag)
CALL addr	2/3	push current PC to stack and branch to addr
RET	2/3	restore PC from stack
TRAP	1/2	push PC to stack and branch to address 001E hex

MNEMONIC	WORDS/ CYCLES	ACTION dma is any indirect addressing mode or the low 7 bits of a data memory address (9 high bits come from DP)
		Control operations
CNFD	1/1	FF00H-FFFFH program memory becomes on chip data memory at 200H-2FFH
CNFP	1/1	200H-2FFH data memory becomes on chip program memory at FF00H-FFFFH
ROVM	1/1	reset overflow mode (7FFFFFFF+1=80000000)
RPTK v	1/1	repeat the next instruction (v-1) times
BIT dma,bn	1/1	copy bit to TC flag, bn is 1s complement of bits 0-15

BITT dma	1/1	copy bit to TC flag, 1s complement of bits 0-15 in T register
LST dma	1/1	read status reg 0
LST1 dma	1/1	read status reg 1
NOP	1/1	No operation
POP	1/1	Pop top of stack to ACC 0-15
POPD dma	1/1	
SOVM	1/1	set overflow mode (7FFFFFFF+1=7FFFFFFF)

APPENDIX 5.2: 256 POINT RADIX 2 FFT FOR THE TMS320C25/C50

```
;radix-2 TMS320C25 fft program
N               equ 256                 ;no. of points
M               equ 8   ;256 = 2**8

                include 'h32025.inc'
                origin 0
scale           equ noshift

                ramaddr 60h             ;1st page of on-chip ram
n2              ramspace 1
tr              ramspace 1
ti              ramspace 1
sina            ramspace 1              ;twiddle factor
cosa            ramspace 1              ;twiddle factor
E               ramspace 1              ;table step
A               ramspace 1
Nt              ramspace 1              ;size of current transform
Ntd2            ramspace 1              ;currrent transform size / 2
temp            ramspace 1

                ramaddr 400h            ;start of off-chip ram
ES              equ 2                   ;element size: real+imaginary
x               ramspace ES * N         ;stored real,imag,real,imag....
;
;
input           equ 0                   ;input port
outps           equ 0                   ;power spectrum output port
outre           equ 1                   ;real output port
outim           equ 2                   ;imaginary output port

                ldpk h.n2               ;set data page pointer
                lrlk ar3,x
                larp ar3
                lrlk ar1,N - 1          ;counter
                lack 0
inloop          in *+,input             ;read real input data
                sacl *+,noshift,ar1     ;imaginary = 0
                banz inloop,*-,ar3
;
;
fftstart        lalk ES * N,shift15     ;starting FFT array size
                sach l.Nt,shift1        ;Nt = ES * N
                sach l.Ntd2,noshift     ;Ntd2 = Nt / 2
                lark ar7,M-1            ;ar7 = k counter
                lack 1                  ;1 table step =
                sacl l.E,noshift        ;2 pi/N
kloop           lack 0
                sacl l.A,noshift        ;A = 0
```

```
                sacl l.n2,noshift       ;n2 = 0
n2loop          lalk sintab,noshift
                add l.A,noshift
                tblr l.sina             ;sina = sin(A)
                addk N / 4              ;90 degrees shift to cos
                tblr l.cosa             ;cosa = cos(B)
                lac l.A,noshift
                add l.E,noshift
                sacl l.A,noshift        ;a = a + e
                lalk x,noshift          ;load x[] pointers
                add l.n2,noshift            sacl l.temp,noshift
                lar ar6,l.temp          ;ar6 = pi = &x[n2]
                add l.Ntd2,noshift
                sacl l.temp,noshift
                lar ar5,l.temp          ;ar5 = pl = &x[n2+Ntd2]
iloop           lar ar0,l.Nt
                larp ar6
                lac *,shift15,ar5
                sub *,shift15,ar6
                sach l.tr,scale         ;tr = *pi - *pl
                lac *,shift15,ar5
                add *+,shift15,ar6
                sach *+,scale           ;*pi = *pi + *pl++
                lac *,shift15,ar5
                sub *,shift15,ar6
                sach l.ti,scale         ;ti = *pi - *pl
                lac *,shift15,ar5
                add *,shift15,ar6
                sach *-,scale,ar5  ;*pi-- = *pi + *pl
                lt l.cosa
                mpy l.ti
                pac
                lt l.sina
                mpy l.tr
                spac
                sach *-,shift1     ;*pl-- = cosa * ti - sina * tr
                lt l.cosa
                mpy l.tr
                pac
                lt l.sina
                mpy l.ti
                apac
                sach *,shift1,ar6  ;*pl = cosa * tr + sina * ti
;
iend            mar *0+,ar5        ;pi = pi + Nt
                mar *0+            ;pl = pl + Nt
                lrlk ar0,x + ES*N ;ar0 = &x[0] + 2*N
                cmpr lt            ;while(pi < (&x[0] + ES * N))
                bbnz iloop         ;iloop if yes
```

```
;
                 lac l.n2,noshift
                 addk ES
                 sacl l.n2,noshift    ;n2 = n2 + ES
                 sub l.Ntd2,noshift ;n2 - Ntd2
n2end            blz n2loop,*-,ar7   ;while(n2 < Ndt2),ar7 = k counter
;
                 lac l.Ntd2,shift15
                 sach l.Nt,shift1     ;Nt = Ntd2
                 sach l.Ntd2,noshift  ;Ntd2 = Ntd2/2 split lengths
                 lac l.E,shift1       ;e = 2 * e
                 sacl l.E,noshift     ;double twiddle factor angle
kend             banz kloop,*-
;
bitrev                                ;shuffle coefficients and output
                                      ;power spectrum
                 lrlk ar1,x           ;ar1 -> x
                 larp ar1
                 lrlk ar0,N           ;ar0=array size/2
                 lrlk ar2,N-1         ;ar2=loop counter
outloop          out *,outre          ;real out
                 sqra *+              ;P=real**2            out *,outim    ;imag
out
                 zac
                 sqra *-              ;acch=real**2,P=imag**2
                 apac                 ;acch = r**2 + i**2
                 sach l.temp,shift1
                 out l.temp,outps   ;power spectrum out
                 mar *br0+,ar2
                 banz outloop,*-,ar1
finish           b finish
;
sintab           ; sine table 256 points/cycle, 32767 == 1.0
                 data 0
                 data 803
                 data 1605
                 data 2407
                 data 3207
                 data 4005
                 data 4801
                 data 5594
                 data 6384
                 data 7170
                 data 7951
                 data 8728
                 data 9499
                 data 10265
                 data 11024
                 data 11777
```

```
            data 12523
            data 13261
            data 13991
            data 14713
            data 15426
            data 16130
            data 16824
            data 17507
            data 18181
            data 18843
            data 19494
            data 20133
            data 20760
            data 21375
            data 21976
            data 22565
            data 23140
            data 23701
            data 24247
            data 24779
            data 25297
            data 25799
            data 26285
            data 26755
            data 27210
            data 27648
            data 28069
            data 28474
            data 28861
            data 29231
            data 29583
            data 29918
            data 30234
            data 30533
            data 30813
            data 31074
            data 31317
            data 31540
            data 31745
            data 31931
            data 32097
            data 32244
            data 32372
            data 32480
            data 32569
            data 32638
            data 32687

costab              ;cosine table starts here
```

```
data 32717
data 32726
data 32717
data 32687
data 32638
data 32569
data 32481
data 32373
data 32245
data 32098
data 31932
data 31747
data 31542
data 31318
data 31076
data 30815
data 30535
data 30237
data 29921
data 29586
data 29234
data 28864
data 28477
data 28073
data 27651
data 27214
data 26759
data 26289
data 25803
data 25301
data 24784
data 24252
data 23705
data 23144
data 22570
data 21981
data 21380
data 20765
data 20138
data 19499
data 18848
data 18186
data 17513
data 16829
data 16135
data 15432
data 14719
data 13997
data 13267
```

```
data 12529
data 11783
data 11030
data 10271
data 9505
data 8734
data 7957
data 7176
data 6390
data 5601
data 4808
data 4012
data 3214
data 2413
data 1612
data 809
data 6
data -796
data -1599
data -2400
data -3201
data -3999
data -4795
data -5588
data -6377
data -7163
data -7945
data -8721
data -9493
data -10258
data -11018
data -11771
data -12517
data -13255
data -13985
data -14707
data -15420
data -16124
data -16818
data -17502
data -18175
data -18838
data -19489
data -20128
data -20755
data -21370
data -21972
data -22560
data -23135
```

```
data -23696
data -24243
data -24775
data -25292
data -25794
data -26281
data -26752
data -27206
data -27644
data -28066
data -28471
data -28858
data -29228
data -29581
data -29915
data -30232
data -30530
data -30810
data -31072
data -31315
data -31539
data -31743
data -31929
data -32096
data -32243
data -32371
data -32479
data -32568
data -32637
data -32687
data -32716
data -32726
data -32717
```

APPENDIX 5.3: GOERTZEL'S ALGORITHM FOR THE TMS320C25/C50

```
;Goertzel's algorithm performed on array x[N]
; bk = e**-j(2*pi*k/N) = cos(2*pi*k/N) - j sin(2*pi*k/N)
; real part of bk = ak/2 = cos(2*pi*k/N)
; imaginary part of bk = sin(2*pi*k/N)*32767
;
                ramaddr 60h                 ;1st page of on-chip ram
Sknm1           ramspace 1                  ;Sk(n-1)
Sknm2           ramspace 1                  ;Sk(n-2)
hak             ramspace 1                  ;ak/2 cos(2*pi*k/N)*32767
bki             ramspace 1                  ;store bk imaginary part here
ykreal          ramspace 1                  ;real result
ykimag          ramspace 1                  ;imaginary result
shift           ramspace 1                  ;store (14-shift) here
                ramaddr 200h                ;x data in on-chip ram
x               ramspace   N                ;x(0) to x(N-1)
;
;
gzel            lrlk ar3,x                  ;*ar3 = x[0]
                lrlk ar1,(N/2-1)            ;loop counter
                zac
                sacl l.Sknm1,noshift        ;Sknm1 = 0
                sacl l.Sknm2,noshift        ;Sknm2 = 0
gloop           lt l.shift                  ;scaling factor
                lact *+                     ;[*px++]-2
                lt l.hak                    ;T = ak/2
                mpy l.Sknm1                 ;P = [ak * Sknm1]-2
                apac                        ;[*px++ + ak*Sknm1]-2
                sub l.Sknm2,(16-shift2)     ; - [Sknm2]-2
                sach l.Sknm2,shift2         ;Skn -> Sknm2
                lt l.shift                  ;scaling factor
                lact *+,ar1                 ;[*px++]-2, load loop counter
                lt l.hak                    ;T = ak/2
                mpy l.Sknm2                 ;P = [ak * Sknm2]-2
                apac                        ;[*px + ak*Sknm2]-2
                sub l.Sknm1,(16-shift2)     ;[Skn]-2
                sach l.Sknm1,shift2         ;Skn -> Sknm1 (next Sknm1)
                banz gloop,*-,ar3
eqn2            mpy l.Sknm1                 ;P=[Sknm1*ak/2]-1 = bkr*Sknm1
                pac                         ;[Sknm1 * bkr]-1
                sub l.Sknm2,(16-shift1)     ;- [Sknm2]-1
                sach l.ykreal,shift1        ; real result
                lt l.bki
                mpy l.Sknm1                 ;P = [bki * Sknm1]-1
                pac                         ;[y(k) imaginary]-1
                sach l.ykimag,shift1        ;imaginary result
gzelret         ret
```

6

FIR filter design methods

6.0   INTRODUCTION

The transfer function of a digital system is specified in terms of the coefficients a_m and b_m as shown in Eq. (1.25). In practice, and particularly for filter designs, these coefficients are normally obtained from a specified frequency response. Depending on the coefficients a_m we have the infinite impulse response (IIR) systems ($a_m \neq 0$) and the finite impulse response (FIR) systems ($a_m = 0$) as illustrated in Fig. 1.4. Their properties are different, and so are their design methods.

In Section 6.1, FIR filters are classified by their specific features. The following Section 6.2 introduces Fourier series and window shaping as the fundamental FIR design methods. An extension to other design methods is given in Section 6.3. More complete designs are given in Section 6.4, for all four filtering types with the use of Matlab programs. The final Section 6.5 discusses the implementation of the multirate filter, from Section 6.4, on the TMS320C25/C50. It produces results in very good agreement with the Matlab result.

6.1   SPECIAL FEATURES OF FIR FILTERS

One of the main characteristics of FIR filters is that they can be designed to have a linear phase response. This is achieved under the condition that the finite impulse (or unit-sample) response is symmetrical about its midpoint. To show this property, consider an FIR filter of length N whose frequency response function, Eq. (1.26) with $a_m = 0$, is given by

$$H(\omega) = \sum_{n=0}^{N-1} h(n) e^{-j\omega nT} \quad , \qquad (6.1)$$

where h(n) are values of the unit-sample response of the filter. In Chapter 1, we have used notation b_m for the filter coefficients. This is changed here to $b_n = h(n)$ since it suits the topics in this chapter and the next. The length of the FIR filter is chosen as N, instead of N+1, in order to conform with the convention in the technical literature.

There are two types of symmetry: **even**, given by

$$h(n) = h(N-1-n) \quad , \tag{6.2}$$

and **odd**, given by

$$h(n) = - h(N-1-n) \quad . \tag{6.3}$$

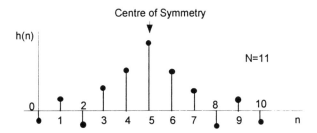

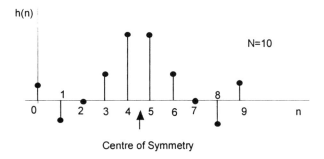

Fig. 6.1: Unit sample response for FIR filters with even symmetry

We illustrate here the case of even symmetry for both the odd and even length filters. If N is odd, (N-1)/2 corresponds to the centre of symmetry of h(n), while if N is even there is no central coefficient, Fig. 6.1.

It is convenient to express the frequency response as

$$H(\omega) = A(\omega) e^{j\theta(\omega)} \quad , \tag{6.4}$$

where $A(\omega)$ is a real-valued amplitude function which may be positive or negative, so that the phase $\theta(\omega)$ is a continuous function of frequency [25].

For **even symmetry** filters, we have $\theta(\omega) = -\omega T(N-1)/2$, and the amplitude function, for **N odd**,

$$A(\omega) = h\left(\frac{N-1}{2}\right) + \sum_{n=0}^{(N-3)/2} 2h(n)\cos\left[\left(\frac{N-1}{2} - n\right)\omega T\right] \qquad (6.5.1)$$

and for **N even** :

$$A(\omega) = \sum_{n=0}^{(N/2)-1} 2h(n)\cos\left[\left(\frac{N-1}{2} - n\right)\omega T\right] \qquad . \qquad (6.5.2)$$

For **odd symmetry**, $\theta(\omega) = \pi/2 - \omega(N-1)/2$, and the amplitude function is, for **N odd**,

$$A(\omega) = \sum_{n=0}^{(N-3)/2} 2h(n)\sin\left[\left(\frac{N-1}{2} - n\right)\omega T\right] \qquad (6.5.3)$$

and for **N even** :

$$A(\omega) = \sum_{n=0}^{(N/2)-1} 2h(n)\sin\left[\left(\frac{N-1}{2} - n\right)\omega T\right] \qquad . \qquad (6.5.4)$$

Note : For odd symmetry, and N odd, $h\left[(N-1)/2\right] = 0$. [27 , p.339].

The filters given by Eqs. (6.5.1) to (6.5.4) are often referred to as Type **1** to Type **4** respectively. It is important to be aware of their specific features :

Types **3** and **4** have A(0) = 0 and hence are not suitable for lowpass filters. Types **2** and **3** have A(π) = 0, hence are not suitable for highpass filters. Also, in addition to linear phase, Types **3** and **4** have constant 90° phase shift which is desirable for a differentiator or a Hilbert transformer [25]. Therefore, Type **1** is generally the most used one for a FIR filter design.

One can determine the coefficients of a FIR filter from samples of the magnitude response in Eq. (6.5.1) or the others. For example, from Eq. (6.5.1) we have

$$\sum_{n=0}^{(N-1)/2} c_{kn} h(n) = |A(\omega_k)| \qquad , \tag{6.6}$$

where

$$c_{kn} = 2 \cos \omega_k \left(\frac{N-1}{2} - n \right) T \tag{6.7}$$

$$c_{kn} = 1, \qquad n = \frac{N-1}{2}, \text{ all } k$$
$$\omega_k = \frac{2\pi}{N} k, \qquad k = 0, 1 \dots, \frac{N-1}{2} \tag{6.8}$$

For a specification, we have $|A(\omega_k)|$ at a discrete set of frequencies $\{\omega_k\}$. Expanding Eq. (6.6) over n, gives a set of linear equations, whose solutions yield the filter coefficients $\{h(n)\}$. Such a design method is acceptable for small filters, but for longer filters one would have to use a digital computer to provide the solution of the linear equations. More detail can be found in [25, 27].

6.2   DESIGN BASED ON FOURIER SERIES AND WINDOWS

This method is often used in FIR design, because it relates $H(\omega)$ and $h(n)$ in a relatively simple way. Since $H(\omega)$ is periodic, its Fourier series gives the filter coefficients $h(n)$ directly.

The desired (ideal) frequency responses for the lowpass, highpass, bandpass and bandstop filters are shown in Fig. 6.2. In these cases, due to the simple form of the function $H(\omega)$, it is easy to obtain the coefficients of the impulse response by evaluating directly the coefficients of the Fourier series using Eq. (1.6) or Eq. (4.2), written here for the impulse response $h(n)$ as

$$h(n) = \frac{1}{2\pi} \int_{-\pi}^{\pi} H(\theta) e^{jn\theta} d\theta \qquad . \tag{6.9}$$

In this equation, and in Fig.6.2, we use the variable

$$\theta = \omega T = 2\pi f/f_s \text{ (radians)} \qquad , \tag{6.10}$$

where $f_s = 1/T$ is the sampling frequency in Hz. (In technical literature, it is often taken that $T=1$, and hence ω (in radians) is used instead of our θ). The baseband range in Fig. 6.2 is given by $0 < \theta < \pi$; this corresponds to $0 < f < f_s/2$, where $f_s/2$ is the half sampling frequency.

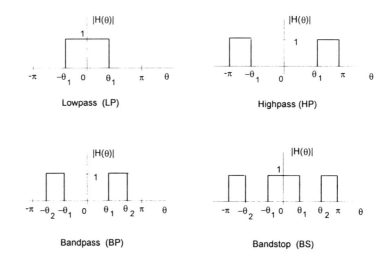

Fig. 6.2: Four types of ideal filter response

The impulse responses of the filters in Fig. 6.2 are summarised in Table 6.1 for an odd number of coefficients and in Table 6.2 for even number of coefficients. Similar expressions can be found in [7] for a differentiator and Hilbert transformer.

Assume that we want to design a lowpass filter with the cutoff $\theta_1 = 0.3\pi$. The actual value of the cutoff frequency f_1 can be obtained as follows.

$$\theta_1 = 2\pi \frac{f_1}{f_s} = 0.3\pi$$

$$f_1 = 0.15 f_s$$

(6.11)

i.e it is controlled by the sampling frequency f_s.

Filter type	Impulse response $\quad n = \pm 1, \pm 2,...., \pm(N-1)/2$
Lowpass	$h(0) = \dfrac{\theta_1}{\pi}, h(n) = \sin(n\theta_1)/n\pi$
Highpass	$h(0) = 1 - \dfrac{\theta_1}{\pi}, h(n) = -\sin(n\theta_1)/n\pi$
Bandpass	$h(0) = \dfrac{\theta_2 - \theta_1}{\pi}, h(n) = \dfrac{1}{n\pi}\left(\sin(n\theta_2) - \sin(n\theta_1)\right)$
Bandstop	$h(0) = 1 - \left(\dfrac{\theta_2 - \theta_1}{\pi}\right), h(n) = \dfrac{1}{n\pi}\left(\sin(n\theta_1) - \sin(n\theta_2)\right)$

Table 6.1 (Impulse response for **odd** number of coefficients)

For the ideal filters of Fig. 6.2 the transition from the passband to the stopband is infinitely steep (i.e discontinuous). For such transitions the impulse response is infinitely long which is not practical. Therefore, the h(n) series must be truncated as indicated in Table 6.1 and 6.2 by N. This results in a gradual transition from the passband to the stopband, and it also produces ripple in the stopband which reduces the stopband attenuation.

Filter type	Impulse response $\quad n = 0, \pm 1,..., \pm\left(\dfrac{N}{2}-1\right), \dfrac{N}{2}$
Lowpass	$h(n) = \left[\sin\left(n-\dfrac{1}{2}\right)\theta_1\right]\Big/\left(n-\dfrac{1}{2}\right)\pi$
Highpass	$h(n) = \left[\sin\left(n-\dfrac{1}{2}\right)\pi - \sin\left(n-\dfrac{1}{2}\right)\theta_1\right]\Big/\left(n-\dfrac{1}{2}\right)\pi$
Bandpass	$h(n) = \left[\sin\left(n-\dfrac{1}{2}\right)\theta_2 - \sin\left(n-\dfrac{1}{2}\right)\theta_1\right]\Big/\left(n-\dfrac{1}{2}\right)\pi$
Bandstop	$h(n) = \dfrac{\sin\left(n-\dfrac{1}{2}\right)\pi - \sin\left(n-\dfrac{1}{2}\right)\theta_2 + \sin\left(n-\dfrac{1}{2}\right)\theta_1}{\left(n-\dfrac{1}{2}\right)\pi}$

Table 6.2 (Impulse response for **even** number of coefficients)

The case of the simple truncation corresponds to multiplication of h(n) by a rectangular window:

$$h_1(n) = h(n) \cdot w_R(n) \quad , \tag{6.12}$$

where

$$w_R(n) = \begin{cases} 1, & |n| < \dfrac{N-1}{2} \\ 0, & \text{otherwise} \end{cases}$$

One can improve the results of truncation by using some kind of window shaping. There are various windows, but we quote here only the generalised cosine windows given as

$$w(n) = a + b \cos\left(\frac{2\pi}{N-1}n\right) + c \cos\left(\frac{4\pi}{N-1}n\right) \tag{6.13}$$

$$0 \le |n| \le (N-1)/2 \quad .$$

The values of constants defining some well know windows, and their performance, are given in Table 6.3.

Window type	a	b	c	Main lobe width (approx)	Minimum stopband attenuation(dB)
Rectangular	1	0	0	$4\pi/N$	-21
Hanning	0.5	0.5	0	$8\pi/N$	-44
Hamming	0.54	0.46	0	$8\pi/N$	-53
Blackman	0.42	0.5	0.08	$12\pi/N$	-74

Table 6.3

Several other windows are available in textbooks such as [25, 27].

Example 6.1

Determine the coefficients of an FIR Filter of length N=13 for the lowpass filter in Fig.6.2 with $\theta_1 = \pi/2$.

Solution: From Table 6.1

$$h(0) = \theta_1/\pi, \; h(n) = \sin(n\theta_1)/n\pi \; ; \; |n| \le \infty \quad .$$

For N = 13 : $|n| \le 6$, we have $h_1(n) = h(n).w_R(n)$

$h_1(0)$	= 0.5		
$h_1(\pm1)$	= 0.318	$h_1(\pm2) = 0$	
$h_1(\pm3)$	= -0.106	$h_1(\pm4) = 0$	
$h_1(\pm5)$	= 0.064	$h_1(\pm6) = 0$	.

Applying now the Hanning (raised-cosine) window:

$$w_H(n) = 0.5\left(1 + \cos\frac{n\pi}{6}\right), |n| \le 6 \quad ,$$

we obtain $h_2(n) = h_1(n) \cdot w_H(n)$ with values

$h_2(0)$	= 0.5		
$h_2(\pm1)$	= 0.296	$h_2(\pm2) = 0$	
$h_2(\pm3)$	= -0.053	$h_2(\pm4) = 0$	
$h_2(\pm5)$	= 0.004	$h_2(\pm6) = 0$	.

To check the frequency response for $\{h_2(n)\}$ we modify Eq. (6.1) as follows:

$$H_2(\omega) = \sum_{n=-(N-1)/2}^{(N-1)/2} h_2(n)e^{-jn\omega T} \quad . \tag{6.14}$$

Introducing also $\theta = \omega T$, we have

$$H_2(\theta) = h_2(0) + 2\sum_{n=1}^{(N-1)/2} h_2(n)\cos n\theta \quad . \tag{6.15}$$

Applying this to our Example, we obtain

$$H2(\theta) = 0.5 + 2\,(0.296\cos\theta - 0.053\cos 3\theta + 0.004\cos 5\theta) \quad .$$

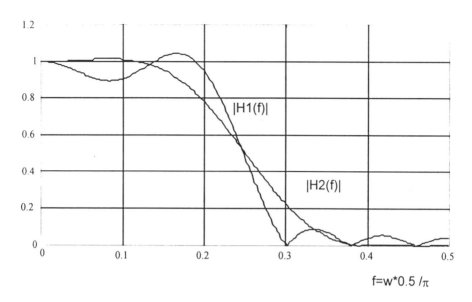

Fig. 6.3: Frequency responses for Example 6.1

Although a relatively simple equation, this is tedious to compute and plot as a graph by hand. But using, for example, the Matlab program given below, we can easily obtain results shown in Fig. 6.3. This program also produces coefficients b1 (for the boxcar, i.e. rectangular window), and b2 (for the Hanning window). These can be compared with h1 and h2, results obtained here by using a pocket calculator. Note that the axes are specified by the variable v, and to superimpose graphs on the same figure hold on and hold off statements have been used.

```
Matlab program for Example 6.1
n = 12;
b1 = fir 1 (n, 0.5, boxcar (n + 1))
[H1, w] = freq z (b1, 1, 256);
v = [0  0.5  0  1.2]; axis (v);
plot (w * 0.5 / pi, abs (H1))
hold on
b2 = fir 1 (n, 0.5 Hanning (n + 1))
[H2, w] = freqz (b2, 1, 256);
plot (w * 0.5 / pi, abs (H2))
grid ; pause
hold off
```

6.3 DESIGN BASED ON FREQUENCY RESPONSE SAMPLING

The design method introduced in the previous section is based on the application of Eq. (6.9) to the desired idealised frequency responses in Fig. 6.2. If these frequency responses are such that no simple mathematical expression can be used to describe them, then the integration in Eq. (6.9) may be difficult to perform. One can use instead a numerical approach in which the desired frequency response is sampled at regular frequency intervals $\omega_k = (2\pi/N)k$. Then, the filter coefficients can be obtained from

$$h(n) = \frac{1}{N} \sum_{k=-(N-1)/2}^{(N-1)/2} H(k)\, e^{j(2\pi/N)kn} \quad , \tag{6.16}$$

where $H(k)$ are magnitude samples of the frequency response as illustrated in Fig. 6.4. This design approach is referred to as the frequency sampling method.

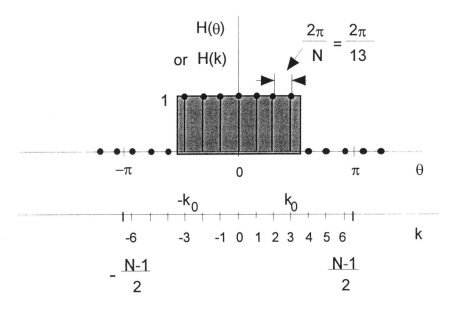

Fig. 6.4: Sampled frequency response for Example 6.2

Since the frequency response is symmetrical around the origin, we have $H(k) = H(-k)$, and the above expression simplifies to

$$h(n) = \frac{1}{N}\left[H(0) + 2 \sum_{k=1}^{(N-1)/2} H(k) \cos\left(\frac{2\pi}{N}kn\right) \right] \quad . \tag{6.17}$$

Example 6.2:

The same as Example 6.1, but using the frequency sampling method. For this case, Eq. (6.17) becomes

$$h(n) = \frac{1}{13}\left[1 + 2 \sum_{k=1}^{3} \cos\left(\frac{2\pi}{13}nk\right) \right], n = 0,\ldots\pm6$$

where H(k) are obtained from Fig 6.4. Results for the coefficients are now:

$$
\begin{aligned}
h(0) &= 0.54 \\
h(\pm1) &= 0.318, & h(\pm2) &= -0.04, \\
h(\pm3) &= -0.108, & h(\pm4) &= 0.04, \\
h(\pm5) &= 0.068, & h(\pm6) &= -0.05,
\end{aligned}
$$

which agree quite well with the values in Example 6.1.

The window method and the frequency sampling method are relatively simple, but they lack the precision which may be required in some specifications.

The alternative is to use the minimax (or equiripple) design based on the minimisation of the maximum error

$$\max | H(\theta) - H'(\theta) | \quad , \tag{6.18}$$

where $H(\theta)$ is a desired frequency response and

$$H'(\theta) = \sum_{n=-M}^{M} h(n)e^{-jn\theta}, \quad \theta = \omega T \tag{6.19}$$

is the actual frequency response. In (6.19) a zero-phase FIR is assumed, i.e. h(n) = - h(n), of duration N=2M+1. The problem of finding H' (θ), i.e. h(n)'s, is called the minimax problem and it is solved by using Chebyshev polynomials. The solutions have equiripple behaviour in their frequency response and hence they are sometimes called equiripple filters. This is quite a complicated theoretical problem, but in practice, solutions are available in computer programs as Remez designs based on the Remez exchange algorithm [25,27].

The filter coefficients (or impulse responses) obtained by the design methods in Sections 6.2 and 6.3 are non-physical (noncausal) because they exist for negative time integers n. The reason for this is the omission of the phase (i.e taking the phase to be zero) in order to simplify design expressions. This is permissible because the resulting filter coefficients are symmetrical which guarantees the linear phase as discussed in Section 6.1. When such a solution is implemented, its peak response will automatically shift from n=0 to n=(N-1)/2, which is in agreement with the linear phase expression in Eq. (6.4) with θ (ω)'s for even and odd symmetry cases.

6.4   FIR FILTER DESIGN EXAMPLES

For higher order filters, or more complicated designs such as Remez, it is necessary to use computer support, as illustrated in the following examples.

Example 6.3
The four designs in Table 6.4 are for N = 21. This is still feasible to do using Table 6.1 and a pocket calculator, but Table 6.4 has been obtained using Matlab.

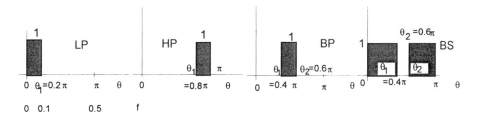

	LP		HP		BP		BS	
	fir1	Remez	fir1	Remez	fir1	Remez	fir1	Remez
h(0) [h(10)]	0.1992	0.2473	0.1992	0.2474	0.237	0.3054	0.8005	0.6946
h(+,-1) [h(9)=h(11)]	0.1821	0.2215	-0.1821	-0.2215	0	0	0	0
h(+,-2) [h(8)=h(12)]	0.1375	0.1538	0.1375	0.1538	-0.2022	-0.2586	0.1708	0.2586
h(+,-3) [h(7)=h(13)]	0.0814	0.0694	-0.0814	-0.0694	0	0	0	0
h(+,-4) [h(6)=h(14)]	0.0318	-0.003	0.0318	-0.003	0.1224	0.1464	-0.1033	-0.1464
h(+,-5) [h(5)=h(15)]	0	-0.0437	0	0.0437	0	0	0	0
h(+,-6) [h(4)=h(16)]	-0.0124	-0.0488	-0.0124	-0.0488	-0.0476	-0.0328	0.0402	0.0328
h(+,-7) [h(3)=h(17)]	-0.0116	-0.0294	0.0116	0.0294	0	0	0	0
h(+,-8) [h(2)=h(18)]	-0.0063	-0.0035	-0.0063	-0.0034	0.0093	-0.0306	-0.0079	0.0306
h(+,-9) [h(1)=h(19)]	-0.0021	0.0141	0.0021	-0.0141	0	0	0	0
h(+,-10) [h(0)=h(20)]	0	0.0378	0	0.0378	0	0.0505	0	-0.0505

Table 6.4: Four FIR designs (N = 21)

The first column is given in terms of h(0), h(±1), …,h(±10). The second column refers to coefficients centred at h(10), as discussed at the end of the previous section. As indicated in Table 6.4, Matlab programs fir1 and Remez have been used to obtain these results. The program fir1(n, w_n) produces a row vector b containing the (n+1) coefficients of the lowpass (LP) filter with cutoff frequency w_n Hamming windowed.

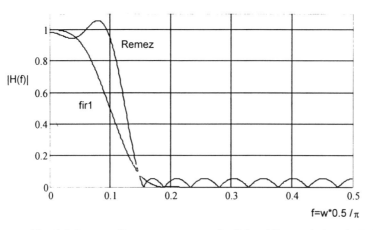

Fig. 6.5: Lowpass frequency responses for fir1 and Remez designs in Table 6.4

The program b = Remez (n, f, m) is based on the equiripple design, with the frequency magnitude response matching given by vectors f and m. These programs use additional steps such as [h, w] and plot for the calculation and plotting of the frequency response (magnitude). This is done for frequencies f = $\theta/2\pi T$, with T=1, so that the range θ = 0 to π is represented by f = 0 to 0.5. Results for the LP case are shown in Fig. 6.5. Similar results are obtained for the other filters in Table 6.4. As mentioned earlier these results can be obtained using the appropriate expressions in Table 6.1 and a pocket calculator. For example, for the lowpass case we have h(0) = 0.1π, and

$$h(n) = \left(\sin n\theta_1\right)/n\pi, \quad 0 \le |n| \le 10 \qquad . \tag{6.20}$$

The Hamming window, from Table 6.2, for this case is

$$h(n) = 0.54 + 0.46 \cos(\pi/10) n, \, 0 \le |n| \le 10 \qquad . \tag{6.21}$$

The multiplications (h·w) produce windowed coefficients. For these results to agree with Matlab figures in Table 6.4, we need to divide (h·w) values by the gain g, which produces b = (h·w)/g. (Do not confuse (h·w) in the above with [h,w] in the Matlab program. The (h·w) here refers to filter coefficients and the window, while the Matlab program h and w refer to the frequency response (h) and frequencies (w)). The **gain g is the peak value of the frequency response of a filter** and it is given by

$$g = \left| \sum_{i=0}^{N-1} (h \cdot w)_i \, z^{-i} \right| \tag{6.22}$$

at $z = 1$ (for LP), at $z = -1$ (for HP), at

$$z = e^{j\pi(f_1+f_2)} \tag{6.23}$$

(for BP), and at $z = 1$ (for BS). For BP, in Table 6.4, $f_1+f_2 = (\theta_1 + \theta_2)/2\pi T = 0.5$ since $T = 1$ for the sampling frequency $f_s = 1$Hz.

Matlab program for Table 6.4
```
b = fir 1 (20, 0.2)
[h, w] = freq z (b, 1, 512);
v = [0   0.5   0   1.1]; axis (v);
plot (w * 0.5 / pi, abs (h))
hold
f = [0   0.2   0.3   1];
m = [1   1   0   0];
b = Remez (20, f, m)
[h, w] = freq z (b, 1, 512);
plot(w*0.5/pi,abs(h);'w')
grid ; pause
hold off
```

The above program is for the LP filter. For other filters the changes are :

HP   :   b = fir 1 (20, 0.8, 'high')
                f = [0   0.7   0.8   1]; m = [0   0   1   1];

BP   :   b = fir 1 (20, [0.4   0.6])
                f = [0   0.3   0.4   0.6   0.7   1]; m = [0   0   1   1   0   0];

BS   :   b = fir 1 (20, [0.4   0.6], 'stop')
                f = [0   0.3   0.4   0.6   0.7   1]; m = [1   1   0   0   1   1];

In Table 6.4, filters are specified as ideal without a transition band from the passband to stopband. In practice, filters are often specified in a more realistic way by using a tolerance scheme, Fig. 6.6. There are three regions specified within the frequency range 0 to F/2, where F is the sampling frequency. The passband frequency range (0 to f_p) has the tolerance specified by $\pm\delta_p$, the stopband (f_s to F/2) has the tolerance given by δ_s. The transition range, between frequencies f_p and f_s, allows the frequency response to fall from the value of one to at least δ_s.

The filter order (N) required to satisfy the specification in Fig. 6.6 can be estimated from

$$N = \frac{-10 \log\left(\delta_p \delta_s\right) - 13}{14 \, \Delta f/F} + 1 \quad , \tag{6.24}$$

where $\Delta f = fs - fp$ [11, 25]   .

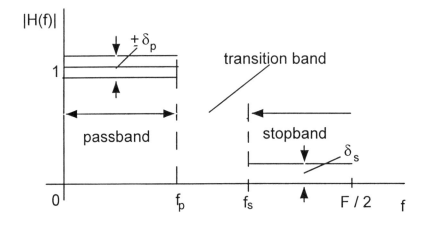

Fig. 6.6: Frequency response tolerance scheme for practical FIR lowpass filters

Example 6.4

For illustration we consider design of a lowpass narrowband filter specified as follows :

$$f_p = 80 \, \text{Hz}, \quad f_s = 100 \text{Hz}, \quad F = 20 \text{kHz}, \quad \delta_p = 0.1, \quad \delta_s = 0.03 \, .$$

Three solutions are considered :
    (i) direct, single rate FIR ;
    (ii) one stage decimator/interpolator and
    (iii) as (ii) but with added intermediate filter.
Cases (ii) and (iii) are multirate designs discussed in Chapter 1, Section 1.6, particularly Figs. 1.13 and 1.14.

Solutions

(i) For the direct implementation, the required filter order is

$$N = \frac{-10\log\left(3 \times 10^{-3}\right) - 13}{14\left(20/20000\right)} + 1 = 874$$

The required DSP machine cycle rate for this filter would be F (S + N) cycles/s, where we perform multiply-accumulate operations at a rate of one per machine cycle after a set-up time of S cycles. This gives 17.68 x 10^6 machine-cycles/s, if we take S = 10.

(ii) As in Fig 1.13, with I = D, and the predecimating filter, Fig. 6.7(a) is allowed to have aliasing on the transition band, which is expressed by

$$\frac{F}{D} - f_p = f_s .$$

Substituting for f_p, f_s, and F as given in the specification, we obtain D = 111.11. We take the next integer value and use D = 112 which changes fs to 98.6 Hz. The required order of the filter, from (6.24), for this case is N = 1168. This is the storage requirement for both filter H1 and H2 in Fig. 1.13.

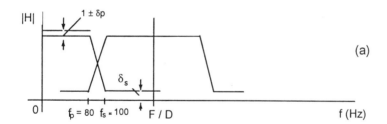

(a)

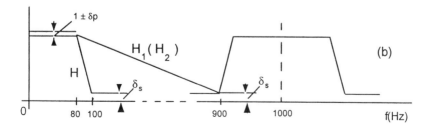

(b)

Figure 6.7: Frequency spectra for decimation/interpolation methods of design

The required machine cycle rate here is

$$2\frac{F}{D}(S+N) = 420714 \qquad \text{machine-cycles/s.}$$

Since the anti-aliasing and anti-imaging coefficients are identical, they need to be stored only once.

In this case, the computational requirements are reduced by a factor of approximately 40, but the storage requirements have doubled.

(iii) As in Fig. 1.14 with $I = D$: the intermediate filter gives us more flexibility and a more efficient computation load. The decimation ratio, D, can now be made less than 112, with the intermediate filter helping to satisfy the system filter specification. We consider the case with $D = I = 20$, for which the spectrum after decimation is as shown in Fig. 6.7(b).

The passband ripple for each filter is 0.033 because the overall ripple 0.1 is to be met by the three filters in cascade. On the other hand, the stopband ripple is taken to be δs in all stages (Note: The passband ripple of ± 0.1 means possible variation from 0.9 to 1.1, while the stopband ripple of 0.03 means variation from 0 to 0.03. The overall response is the product of the three cascaded units).

The filter H1(f) is now allowed to have the transition band up to 900 Hz, hence $\Delta f = 820$ Hz. For this case Eq. (6.24) gives the filter order $N_1 = 31$. The specification for the intermediate filter, H(f), is the same as for the original, but the sampling frequency is $F/D = 1000$ Hz, and $\delta p = 0.033$. The required filter order, from Eq. (6.24) is now $N_2 = 62$. The total computation requirements are :

$$2\frac{F}{D}(S+N_1) + \frac{F}{D}(S+N_2) \qquad = 154000 \text{ machine-cycles/s,}$$

which is a significant improvement.

The above calculations of computational requirements take no account of some overheads which are essential in a real implementation, especially the movement of data. The manipulations needed to implement practical multirate filters efficiently are described in the rest of this chapter.

Case (iii) can also be explored with the Matlab program given below. The input is an allpass signal x, used for testing as in [10]. It is a pulse $x(0) = -0.9$, $x(1) = 0.19$ and then decaying asymptotically to zero. This signal in the time domain is given in stem (n,x) and its frequency flat spectrum is obtained from plot (f, X1 (1:256)). The decimation and interpolation steps, by factors $D = I = 20$, are performed using decimate and interp programs from [18]. The results of these operations can be observed from stem (td,xd) and stem (ti,xi); the latter shows nicely the intermediate filter impulse response.

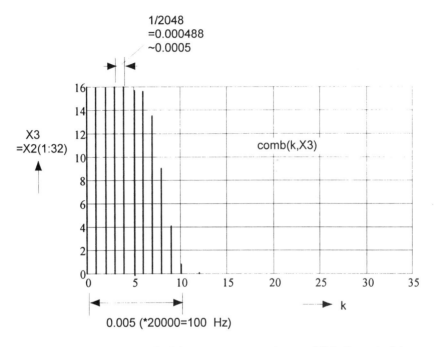

Fig. 6.8: Detail of the output spectrum for case (iii) in Example 6.4

The intermediate filtering is obtained by the statements b = fir 1 (62, 0.16) and y = filter (b,1,xd), as described in [18]. The bandwidth wn = 0.16 is obtained from the passband cutoff 80 Hz normalised to the new sampling frequency 1000 Hz, producing 0.08. (In Matlab design programs, the frequency scale is set to be 0 to 1 where 1 refers to the sampling frequency; therefore 0.08 changes to 0.16). The output spectrum from the interpolator is obtained from plot (f, X2 (1 : 1024)) for a 2048 point FFT. It is appropriate to use this size of FFT in order to see clearly the final result of the expanded spectrum (k=0 to 31 points), as shown in Fig. 6.8. The number of input signal points has also been taken as 2048 in order to avoid spectral leakage which would obscure the output spectrum. The input signal used here is very useful for testing, since it provides a wide band flat spectrum so that filtering results can be clearly seen at the output of the processing chain.

The intermediate filter can easily be 'commented out' by inserting % at the front of the statements for b and y, and also by changing y in xi statement to xd. The output spectrum is then obtained from plot (f,X2 (1 : 1024)), as shown in Fig 6.9. It is also interesting in this case to see stem (ti,xi) = stem (n,x) after the decimation and interpolation operations.

Matlab program for Example 6.4 (case iii)

```
t = 1 : 1 : 2047 ; a = 0.9;
x = [ -a          (1 - a^2) * a.^ (t-1];
n = 0 : 1 : 2047;
stem (n,x) ; grid ; pause ; clf
X = fft (x,512) ; X1 = abs (X) ;
f = (0 : 255) / 512 ;
plot (f, X1(1 : 256)) ; grid ; pause ; clf
t1 = 0 : 1 : 2047 ;
td = decimate (t1,20,31,'fir') ;
xd = decimate (x, 20, 31, 'fir') ;
stem (td, xd) ; grid ; pause ; clf
b = fir 1 (62, 0.16) ;
y = filter (b, 1, xd) ;
ti = interp (td, 20) ;
xi = interp (y, 20) ;
stem (ti, xi) ; grid ; pause ; clf
Xi = fft (xi, 2048) ; X2 = abs (Xi) ;
f = (0 : 1023) / 2048 ;
plot (f , X2 ( 1 : 1024)) ; grid ; pause ; clf
X3 = [X2 (1 : 32)] ; k = 0 : 1 : 31 ;
stem (k, X3) ; grid
```

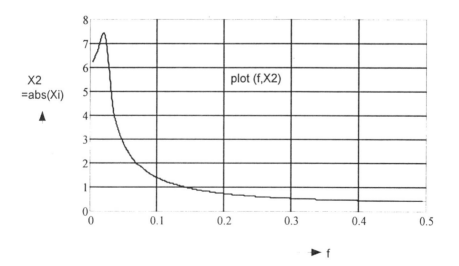

Fig. 6.9: Output spectrum without the intermediate filter

6.5 THE IMPLEMENTATION OF DECIMATION AND INTERPOLATION

The normal implementation of a FIR filter requires the storage of past data x(n) to x(n-N-1) for a filter with N coefficients h(0) to h(N-1). These N data samples are moved as each new sample appears, so that the last N samples are always stored.

coeff-icients	TIME → A	B	C	D	E	F	G	H
h(0)	x(n)	x(n+1)	x(n+2)	x(n+3)	x(n+4)	x(n+5)	x(n+6)	x(n+7)
h(1)	x(n-1)	x(n)	x(n+1)	x(n+2)	x(n+3)	x(n+4)	x(n+5)	x(n+6)
h(2)	x(n-2)	x(n-1)	x(n)	x(n+1)	x(n+2)	x(n+3)	x(n+4)	x(n+5)
h(3)	x(n-3)	x(n-2)	x(n-1)	x(n)	x(n+1)	x(n+2)	x(n+3)	x(n+4)
h(4)	x(n-4)	x(n-3)	x(n-2)	x(n-1)	x(n)	x(n+1)	x(n+2)	x(n+3)
h(5)	x(n-5)	x(n-4)	x(n-3)	x(n-2)	x(n-1)	x(n)	x(n+1)	x(n+2)
h(6)	x(n-6)	x(n-5)	x(n-4)	x(n-3)	x(n-2)	x(n-1)	x(n)	x(n+1)
h(7)	x(n-7)	x(n-6)	x(n-5)	x(n-4)	x(n-3)	x(n-2)	x(n-1)	x(n)
h(8)	x(n-8)	x(n-7)	x(n-6)	x(n-5)	x(n-4)	x(n-3)	x(n-2)	x(n-1)
h(9)	x(n-9)	x(n-8)	x(n-7)	x(n-6)	x(n-5)	x(n-4)	x(n-3)	x(n-2)
h(10)	x(n-10)	x(n-9)	x(n-8)	x(n-7)	x(n-6)	x(n-5)	x(n-4)	x(n-3)
h(11)	x(n-11)	x(n-10)	x(n-9)	x(n-8)	x(n-7)	x(n-6)	x(n-5)	x(n-4)

Figure 6.10: FIR filter, N=12, decimated, D=3 and rearranged to form 3 FIR filters,
N = 4 whose sum is the output
(a) States of FIR filter operating at the full data rate

coeff-icients	TIME → A	B	C	D	E	F	G	H
h(0)	x(n)			x(n+3)			x(n+6)	
h(1)	x(n-1)			x(n+2)			x(n+5)	
h(2)	x(n-2)			x(n+1)			x(n+4)	
h(3)	x(n-3)			x(n)			x(n+3)	
h(4)	x(n-4)			x(n-1)			x(n+2)	
h(5)	x(n-5)			x(n-2)			x(n+1)	
h(6)	x(n-6)			x(n-3)			x(n)	
h(7)	x(n-7)			x(n-4)			x(n-1)	
h(8)	x(n-8)			x(n-5)			x(n-2)	
h(9)	x(n-9)			x(n-6)			x(n-3)	
h(10)	x(n-10)			x(n-7)			x(n-4)	
h(11)	x(n-11)			x(n-8)			x(n-5)	
	=y(n)			=y(n+3)			=y(n+6)	

Figure 6.10: FIR filter, N=12, decimated, D=3 and rearranged to form 3 FIR filters,
N = 4 whose sum is the output
(b) Essential states of filter (a) decimated, D = 3

coeff-icients	TIME →							
	A	B	C	D	E	F	G	H
h(0)	x(n)			x(n+3)			x(n+6)	
h(1)			x(n+2)			x(n+5)		
h(2)		x(n+1)			x(n+4)			
h(3)	x(n-3)			x(n)			x(n+3)	
h(4)			x(n-1)			x(n+2)		
h(5)		x(n-2)			x(n+1)			
h(6)	x(n-6)			x(n-3)			x(n)	
h(7)			x(n-4)			x(n-1)		
h(8)		x(n-5)			x(n-2)			
h(9)	x(n-9)			x(n-6)			x(n-3)	
h(10)			x(n-7)			x(n-4)		
h(11)		x(n-8)			x(n-5)			
	=p0	=p2	=p1	=p0	=p2	=p1	=p0	=p2
				y(n+3)=P0+P1+P2			y(n+6)=P0+P1+P2	

Figure 6.10: FIR filter, N=12, decimated, D=3 and rearranged to form 3 FIR filters,
N = 4 whose sum is the output
(c) Decimated filter with partial filters computed at earliest opportunity

As we have seen in Section 3.2 et. seq. all DSPs possess architectural features that allow the multiplication of stored data samples by coefficients, with summation and moving of the data samples at a rate approaching one operation (multiply + add + move) per machine cycle. The passage of data through a filter with N=12 is shown in Fig. 6.10a, showing the stored samples for the 8 time-slots A to H. We shall now follow this filter by a decimation stage, (refer to Fig. 1.13). If we decimate the output of the filter of Fig. 6.10a (D = 3), we have the situation shown in Fig. 6.10b. Note that although we have removed 2/3 of the multiplications and additions, we have not reduced the need to shift the input data. Thus in Fig. 6.10b, the array of samples x( ) has to be shifted during slots B and C to be in the correct locations for the computation of sample 2. In many DSPs, this would give hardly any improvement in speed as shifting the data would take just as long as computing the missing filter elements, (note though that a circular buffer capability, see 2.5.2, may ameliorate this situation).

6.5.1   Decimation implementation

We may note however in Fig. 6.10b that not all the samples are multiplied by all the coefficients, so that x(n) is only used with h0, h3, h6 etc. x(n+1) is only used with h2, h5, h8 and so on. Therefore we can decide to compute the components of the decimated filter output in D stages, shown in Fig. 6.10c. Thus in preparation for decimated sample y(n+3), we compute 1/3 of the multiply-accumulates (P2), using x(n+1), x(n-2), x(n-5),x(n-8) during time slot B, x(n+2) etc. at time C (P2) and finally x(n+3) etc. It can be seen that we now have three FIR filter structures, P2, P1 and P0 whose outputs can be summed to get the desired filtered, decimated output. The coefficients and the data samples can be reordered in the memory of the DSP as shown in Fig. 6.11 to make use of the DSP capability to perform the

coeff-icients	TIME →							
	A	B	C	D	E	F	G	H
h(0)	x(n)			x(n+3)			x(n+6)	
h(3)	x(n-3)			x(n)			x(n+3)	
h(6)	x(n-6)			x(n-3)			x(n)	
h(9)	x(n-9)			x(n-6)			x(n+3)	
h(1)			x(n+2)			x(n+5)		
h(4)			x(n-1)			x(n+2)		
h(7)			x(n-4)			x(n-1)		
h(10)			x(n-7)			x(n-4)		
h(2)		x(n+1)			x(n+4)			x(n+7)
h(5)		x(n-2)			x(n+1)			x(n+4)
h(8)		x(n-5)			x(n-2)			x(n+1)
h(11)		x(n-8)			x(n-5)			x(n-2)
	=p0	=p2	=p1	=p0	=p2	=p1	=p0	=p2
				y(n+3)= P2+ P1+ P0			y(n+6)= P2+ P1+ P0	

Fig. 6.11: Decimated filter coefficients rearranged for standard DSP convolution

multiply, add and data move operations concurrently. It is also clear from Fig. 6.11 that, the calculations for P2, P1 and P0 can be combined into a single convolution as long as the necessary data are in place, e.g. as long as x(n+1), x(n+2), x(n+3) have been stored, the calculation of y(n+3) can be performed on the TMS320C25 with a single repeated MACD instruction, see Section 3.3.2. This implies that D samples can be stored, perhaps in a FIFO buffer until they can be written into the array of samples to be convolved. Note that a restriction for this efficient implementation to be possible is that N/D is an integer.

6.5.2   Interpolation implementation

The interpolation stage of Fig. 1.13 can similarly be implemented by a set of FIR filters. If we take the decimated data stream just shown and interpolate zeros in between the samples in order to restore the original data rate, we get x(n), 0, 0, x(n+3), 0, 0, x(n+6) etc.

coeff-icients	TIME →							
	A	B	C	D	E	F	G	H
h(0)	x(n)	0	0	x(n+3)	0	0	x(n+6)	0
h(1)	0	x(n)	0	0	x(n+3)	0	0	x(n+6)
h(2)	0	0	x(n)	0	0	x(n+3)	0	0
h(3)	x(n-3)	0	0	x(n)	0	0	x(n+3)	0
h(4)	0	x(n-3)	0	0	x(n)	0	0	x(n+3)
h(5)	0	0	x(n-3)	0	0	x(n)	0	0
h(6)	x(n-6)	0	0	x(n-3)	0	0	x(n)	0
h(7)	0	x(n-6)	0	0	x(n-3)	0	0	x(n)
h(8)	0	0	x(n-6)	0	0	x(n-3)	0	0
h(9)	x(n-9)	0	0	x(n-6)	0	0	x(n-3)	0
h(10)	0	x(n-9)	0	0	x(n-6)	0	0	x(n-3)
h(11)	0	0	x(n-9)	0	0	x(n-6)	0	0

Fig. 6.12: Interpolation, followed by a FIR filter (N = 12)

In Fig. 6.12, we show a FIR filter operating at the output sample rate with the state of the storage elements as time progresses. Obviously the multiplications by zero can be eliminated and Fig. 6.13 shows the system rearranged as in the decimation case into a set of D filters, each of length N/D.

coeff-icients	TIME →							
	A	B	C	D	E	F	G	H
h(0)	x(n)			x(n+3)			x(n+6)	
h(3)	x(n-3)			x(n)			x(n+3)	
h(6)	x(n-6)			x(n-3)			x(n)	
h(9)	x(n-9)			x(n-6)			x(n+3)	
h(1)		x(n)			x(n+3)			x(n+6)
h(4)		x(n-3)			x(n)			x(n+3)
h(7)		x(n-6)			x(n-3)			x(n)
h(10)		x(n-9)			x(n-6)			x(n+3)
h(2)			x(n)			x(n+3)		
h(5)			x(n-3)			x(n)		
h(8)			x(n-6)			x(n-3)		
h(11)			x(n-9)			x(n-6)		
	y(n)= P0	y(n+1)= P1	y(n+2)= P2	y(n+3)= P0	y(n+4)= P1	y(n+5)= P2	y(n+5)= P0	y(n+6)= P1

Fig. 6.13: Interpolator (D=3) and filter with coefficients rearranged for standard DSP convolutions

In this case, the three filters cannot be combined into a single convolution as the separate outputs are required, so that interpolation will be slower than decimation in a practical system.

6.5.3 Decimation/interpolation as a polyphase filter

In a filter where K = N/D, an integer, for example: N = 12, D = 3, K = 4 we can describe the decimation followed by filtering as a set of polyphase filters, as shown in Fig. 6.14a and the interpolation in a similar fashion as in Fig. 6.14b, see Proakis & Manolakis [27, p. 776].

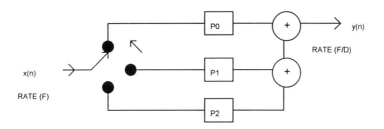

(a) Decimation with polyphase filters

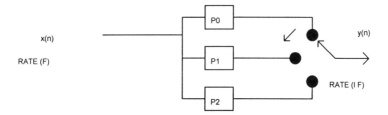

(b) Interpolation with polyphase filters

Fig. 6.14: Polyphase filter interpolation/decimation

Basic equations for the filters are:

$pk(n) = h(nD + k)$ $k = 0,1, ... ,D\text{-}1$ (I=D=3: $k = 0,1,2$)

$pk(n) = h(nI + k)$ $k = 0,1, ... , I\text{-}1$

                                             $n = 0,1, ... ,K\text{-}1$ (K=4: $n = 0,1,2,3$) .

These expand to:

$p0(n) = h(3n)$: $P0 = [\, h(0)\ h(3)\ h(6)\ h(9) \,]$

$p1(n) = h(3n+1)$: $P1 = [\, h(1)\ h(4)\ h(7)\ h(10) \,]$

$p2(n) = h(3n+2)$: $P2 = [\, h(2)\ h(5)\ h(8)\ h(11) \,]$.

We would like to compute the interpolation and decimation samples in chronological order. For the decimation this means the order P2, P1, P0 . Note that D samples need to be input before the decimated output can be computed. The D interpolated outputs can be computed in the order P0, P1, P2 as soon as each (non-zero) decimated value is read. Notice that this 'cancels out' the delay in the decimation stage and means that a multirate system need not produce any more delay than a simple filter.

6.5.4 Execution times for decimation and interpolation

On a DSP, a convolution is normally performed by the repetition of a single instruction, such as MACD on the TMS320C25/C50. When the multiply/accumulate starts repeating, it executes at a rate of one machine cycle per step. Therefore a convolution, such as the decimation filter combining p0, p1 and p2, using N samples takes $(S + N)$ machine cycles per sample. S machine cycles are required to set up address pointers etc. Therefore if the input sample rate is F, and the decimation ratio is D, the DSP can perform

$$\frac{F}{D}(S+N) \quad \text{samples/s.}$$

The interpolation stage must be executed as D convolutions of length N/D, executed at rate F. This gives

$$F(S + \frac{N}{D}) \quad \text{samples/s ;}$$

S is likely to be around 10 machine cycles for the TMS320C25/C50.

6.5.5 Multirate filter implementation

We show here a practical implementation on the TMS320C25/C50 of Example 6.4, given earlier, implementing a LP filter with fp = 80Hz and sampling rate 20 kHz. The scheme implemented is version (iii) with an intermediate filter. In this case N = 64, the decimation ratio D = 16, giving N/D = 4. The intermediate filter also has 64 coefficients. The machine cycles for each stage of the program are given in Fig. 6.15.

STAGE	OPERATION	MACHINE CYCLES
1	Read 16 samples to input array (size 64)	19
2	Decimation: Convolve input array with 64 coefficients scale decimated result → intermediate filter array (size 64)	72
3	Convolve 64 point FIR intermediate filter scale result → output array (size 4)	74

4	Interpolation: Convolve output array → output y(n)
	              Convolve output array → output y(n+1)
	                          etc.
	              Convolve output array → output y(n+15)
		245

Fig. 6.15: Timing of the multirate filter on the TMS320C25

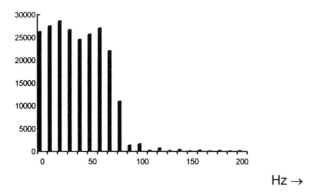

Fig. 6.16: Frequency response of a TMS320C25/C50 implementation of the LP filter (fp = 80 Hz)

Note that these are not theoretical but measured values for a system, which can operate in real time. The 410 machine cycles to process 16 input samples means that a DSP with 100 ns cycle time would only use about 5% of the available processing power. Each convolution of the interpolation stage uses only 4 samples, which means that the setting up overhead for the MACD instruction is large. The efficiency of the interpolation could be speeded up by alternative coding. The measured frequency response is shown in Fig. 6.16, and the program used is shown in Appendix 6.1.

APPENDIX   6.1: TMS320C25/C50 MULTIRATE FIR PROGRAM

(used to produce Fig. 6.16)

```
;**multirate filter for TMS320C25/C50
;DECIMATOR -> FIR -> INTERPOLATOR

                include 'h32025.inc'

;  Note that subroutine 'response' is used for obtaining output data
;  and is removed for execution speed measurements
;  Ports 1 and 2 are used to obtain files of internal data
; and are also commented out here


Nd              equ 64          ;decimator filter length
D               equ 16          ;decimation ratio
Nf              equ 64          ;intermediate filter length
input           equ 0           ;input port
output          equ 0           ;output port

                ramaddr 300h
x0              ramspace Nd+1       ;input samples x(0) to x(Nd-1)
f0              ramspace Nf+1       ;intermediate filter samples
i0              ramspace Nd/D+1             ;interpolation samples
temp            ramspace 1

                ldpk h.temp     ;page pointer
                spm pright6     ;to avoid overflow
        call resinit
                lrlk ar1,x0     ;ar1-> input samples
                lrlk ar2,f0     ;ar2-> intermediate samples
                lrlk ar3,i0     ;ar3-> interpolation samples
                lark ar0,Nd/D   ;ar0=subfilter size
                larp ar1
;DECIMATOR FILTER:
loop            rptk D-1
                in *0+,input    ;get D samples
                mar *-          ;ar1 -> oldest sample
                lack 0          ;acc = 0
                mpyk 0           ;P = 0
                rptk Nd-1
                macd h15,*-     ;decimation filter
                apac            ;last add
                mar *+,ar2      ;ar1 -> 1st data sample ar2 ->
```

```
;INTERMEDIATE FILTER:
decend          sach *,shift1    ;intermediate input

;        out *,1
                lrlk ar2,f0+Nf-1 ;ar2 -> oldest sample
                lack 0       ;acc = 0
                mpyk 0       ;P = 0
                rptk Nf-1
                macd h0f,*-    ;convolution
                apac          ;last add
                mar *+,ar3       ;ar2 -> 1st data sample ar3 ->
;INTERPOLATION FILTER:
firend          sach *,shift4        ;decimation input

;        out *,2
                lrlk ar3,((i0+Nd/D)-1) ;last sample
                mpyk 0       ;P = 0
                zac          ;acc = 0
                rptk (Nd/D-1)
                mac h0,*-
                apac
                sach l.temp,shift5
                out l.temp,output
;        call response
                mar *0+          ;last sample
                mpyk 0       ;P = 0
                zac          ;acc = 0
                rptk (Nd/D-1)
                mac h1,*-
                apac
                sach l.temp,shift5
                out l.temp,output
;        call response
                mar *0+          ;last sample
                mpyk 0       ;P = 0
                zac          ;acc = 0
                rptk (Nd/D-1)
                mac h2,*-
                apac
                sach l.temp,shift5
                out l.temp,output
;        call response
                mar *0+          ;last sample
                mpyk 0       ;P = 0
                zac          ;acc = 0
                rptk (Nd/D-1)
                mac h3,*-
                apac
                sach l.temp,shift5
```

```
                        out l.temp,output
;       call response
                        mar *0+          ;last sample
                        mpyk 0      ;P = 0
                        zac         ;acc = 0
                        rptk (Nd/D-1)
                        mac h4,*-
                        apac
                        sach l.temp,shift5
                        out l.temp,output
;       call response
                        mar *0+          ;last sample
                        mpyk 0      ;P = 0
                        zac         ;acc = 0
                        rptk (Nd/D-1)
                        mac h5,*-
                        apac
                        sach l.temp,shift5
                        out l.temp,output
;       call response
                        mar *0+          ;last sample
                        mpyk 0      ;P = 0
                        zac         ;acc = 0
                        rptk (Nd/D-1)
                        mac h6,*-
                        apac
                        sach l.temp,shift5
                        out l.temp,output
;       call response
                        mar *0+          ;last sample
                        mpyk 0      ;P = 0
                        zac         ;acc = 0
                        rptk (Nd/D-1)
                        mac h7,*-
                        apac
                        sach l.temp,shift5
                        out l.temp,output
;       call response
                        mar *0+          ;last sample
                        mpyk 0      ;P = 0
                        zac         ;acc = 0
                        rptk (Nd/D-1)
                        mac h7,*-
                        apac
                        sach l.temp,shift5
                        out l.temp,output
;       call response
                        mar *0+          ;last sample
                        mpyk 0      ;P = 0
```

```
                zac            ;acc = 0
                rptk (Nd/D-1)
                mac h9,*-
                apac
                sach l.temp,shift5
                out l.temp,output
;      call response
                mar *0+           ;last sample
                mpyk 0         ;P = 0
                zac           ;acc = 0
                rptk (Nd/D-1)
                mac h10,*-
                apac
                sach l.temp,shift5
                out l.temp,output
;      call response
                mar *0+           ;last sample
                mpyk 0         ;P = 0
                zac           ;acc = 0
                rptk (Nd/D-1)
                mac h11,*-
                apac
                sach l.temp,shift5
                out l.temp,output
;      call response
                mar *0+           ;last sample
                mpyk 0         ;P = 0
                zac           ;acc = 0
                rptk (Nd/D-1)
                mac h12,*-
                apac
                sach l.temp,shift5
                out l.temp,output
;      call response
                mar *0+           ;last sample
                mpyk 0         ;P = 0
                zac           ;acc = 0
                rptk (Nd/D-1)
                mac h13,*-
                apac
                sach l.temp,shift5
;      call response
                out l.temp,output
                mar *0+           ;last sample
                mpyk 0         ;P = 0
                zac           ;acc = 0
                rptk (Nd/D-1)
                mac h14,*-
                apac
```

```
                      sach I.temp,shift5
                      out I.temp,output
;            call response
                      mar *0+              ;last sample
                      mpyk 0        ;P = 0
                      zac          ;acc = 0
                      rptk (Nd/D-1)
                      macd h15,*-    ;dmov last time
                      apac
                      sach I.temp,shift5
                      out I.temp,output
;            call response
                      mar *+,ar1 ;ar3 -> i0
                      b loop


             ;FIR coefficients, f/fs=0.004000000, decimation = 16
h15          data   29450 ;h[ 63]
             data   31945 ;h[ 47]
             data   32767 ;h[ 31]
             data   31836 ;h[ 15]
h14          data   29651 ;h[ 62]
             data   32047 ;h[ 46]
             data   32760 ;h[ 30]
             data   31721 ;h[ 14]
h13          data   29847 ;h[ 61]
             data   32142 ;h[ 45]
             data   32746 ;h[ 29]
             data   31599 ;h[ 13]
h12          data   30036 ;h[ 60]
             data   32231 ;h[ 44]
             data   32725 ;h[ 28]
             data   31471 ;h[ 12]
h11          data   30220 ;h[ 59]
             data   32313 ;h[ 43]
             data   32698 ;h[ 27]
             data   31337 ;h[ 11]
h10          data   30398 ;h[ 58]
             data   32388 ;h[ 42]
             data   32663 ;h[ 26]
             data   31196 ;h[ 10]
h9           data   30570 ;h[ 57]
             data   32457 ;h[ 41]
             data   32622 ;h[ 25]
             data   31049 ;h[ 9]
h8           data   30736 ;h[ 56]
             data   32519 ;h[ 40]
             data   32574 ;h[ 24]
             data   30895 ;h[ 8]
h7           data   30895 ;h[ 55]
```

```
        data    32574 ;h[ 39]
        data    32519 ;h[ 23]
        data    30736 ;h[  7]
h6      data    31049 ;h[ 54]
        data    32622 ;h[ 38]
        data    32457 ;h[ 22]
        data    30570 ;h[  6]
h5      data    31196 ;h[ 53]
        data    32663 ;h[ 37]
        data    32388 ;h[ 21]
        data    30398 ;h[  5]
h4      data    31337 ;h[ 52]
        data    32698 ;h[ 36]
        data    32313 ;h[ 20]
        data    30220 ;h[  4]
h3      data    31471 ;h[ 51]
        data    32725 ;h[ 35]
        data    32231 ;h[ 19]
        data    30036 ;h[  3]
h2      data    31599 ;h[ 50]
        data    32746 ;h[ 34]
        data    32142 ;h[ 18]
        data    29847 ;h[  2]
h1      data    31721 ;h[ 49]
        data    32760 ;h[ 33]
        data    32047 ;h[ 17]
        data    29651 ;h[  1]
h0      data    31836 ;h[ 48]
        data    32767 ;h[ 32]
        data    31945 ;h[ 16]
        data    29450 ;h[  0]

h0f     ;FIR coefficients, f/fs=0.064000003
        data     261 ;h[  0]
        data    -798 ;h[  1]
        data   -1799 ;h[  2]
        data   -2572 ;h[  3]
        data   -2977 ;h[  4]
        data   -2919 ;h[  5]
        data   -2372 ;h[  6]
        data   -1387 ;h[  7]
        data     -87  ;h[  8]
        data    1342 ;h[  9]
        data    2681 ;h[ 10]
        data    3701 ;h[ 11]
        data    4206 ;h[ 12]
        data    4058 ;h[ 13]
        data    3209 ;h[ 14]
        data    1713 ;h[ 15]
```

```
data    -265 ;h[ 16]
data    -2473 ;h[ 17]
data    -4583 ;h[ 18]
data    -6241 ;h[ 19]
data    -7106 ;h[ 20]
data    -6893 ;h[ 21]
data    -5420 ;h[ 22]
data    -2634 ;h[ 23]
data     1370 ;h[ 24]
data     6356 ;h[ 25]
data    11955 ;h[ 26]
data    17713 ;h[ 27]
data    23126 ;h[ 28]
data    27706 ;h[ 29]
data    31024 ;h[ 30]
data    32767 ;h[ 31]
data    32767 ;h[ 32]
data    31024 ;h[ 33]
data    27706 ;h[ 34]
data    23126 ;h[ 35]
data    17713 ;h[ 36]
data    11955 ;h[ 37]
data     6356 ;h[ 38]
data     1370 ;h[ 39]
data    -2634 ;h[ 40]
data    -5420 ;h[ 41]
data    -6893 ;h[ 42]
data    -7106 ;h[ 43]
data    -6241 ;h[ 44]
data    -4583 ;h[ 45]
data    -2473 ;h[ 46]
data    -265 ;h[ 47]
data     1713 ;h[ 48]
data     3209 ;h[ 49]
data     4058 ;h[ 50]
data     4206 ;h[ 51]
data     3701 ;h[ 52]
data     2681 ;h[ 53]
data     1342 ;h[ 54]
data     -87 ;h[ 55]
data    -1387 ;h[ 56]
data    -2372 ;h[ 57]
data    -2919 ;h[ 58]
data    -2977 ;h[ 59]
data    -2572 ;h[ 60]
data    -1799 ;h[ 61]
data    -798 ;h[ 62]
data     261 ;h[ 63]
```

```
;computes maximum value in temp -> rmax
;for rsamples calls & outputs to port 15
;resinit initialises
rsamples                 equ 1500  ;sample count
rmax          ramspace 1
rcount        ramspace 1

response      lac l.temp,noshift
              abs
              sub l.rmax,noshift
              blz rlab1
              add l.rmax,noshift
              sacl l.rmax,noshift
rlab1         zals l.rcount
              subk 1
              sacl l.rcount,noshift
              bgz rlab2
              out l.rmax,15
rout          nop
resinit       lalk rsamples,noshift ;initialise system
              sacl l.rcount,noshift
              zac
              sacl l.rmax,noshift
rlab2         ret
```

7

IIR Filter Design

7.0   INTRODUCTION

The linear phase FIR filter designs are based on Eq. (6.9), because it gives the required filter coefficients $\{h(n)\}$ = $\{b_n\}$ directly. This method is not suitable for IIR filters because $\{h(n)\}$ values are not simply related to filter coefficients $\{a_m\}$ and $\{b_m\}$.

This chapter introduces two main IIR filter design techniques in Sections 7.1 and 7.2. They are based on the idea of converting an analogue filter to a digital filter. The transformations from the prototype lowpass filter to other types of filters are given in Section 7.3. The characteristics and design features of commonly used analogue filters in IIR designs, are given in Section 7.4. This is followed, in Section 7.5, by the derivation of IIR filter structures and their corresponding difference equations for computer implementations.

The quantization effects are considered briefly in Section 7.6. The following Section 7.7 deals with four filter designs using the transformations from the prototype lowpass filter, and it also gives Matlab instructions for the designs.

The final Section 7.8 describes DSP implementation with results for some designs in Section 7.7. A complete listing of programs used is given in Appendix 7.2.

7.1   IMPULSE INVARIANCE (II) METHOD

The objective is to design an IIR filter having impulse (or unit-sample) response h(n) which is a sampled version of the impulse response of the analogue filter. The design relationship is summarised in the following:

$$H(s) = \sum_{k=1}^{M} \frac{A_n}{s - s_k} \rightarrow \sum_{k=1}^{M} \frac{A_n T}{1 - e^{s_k T} z^{-1}} = H(z) \qquad , \qquad (7.1)$$

where H(s) represents the analogue filter transfer function expanded into partial fractions form with simple poles at s_k.

It is seen that a pole s_k, in the s-plane, transforms to a pole at $z_k = \exp(s_k T)$ in the z-plane. The (s,z) relationship is $z = e^{sT}$, i.e. the standard z-transform introduced by Eq. (1.14) and illustrated in Fig. 1.3.

As usual, one has to check the frequency response of the designed filter over the frequency range 0 to $f_s/2$, where f_s is the sampling frequency. It will be found that there is an error between $|H(s=j\Omega)|$, the analogue filter response, and $|H(z=\exp(j\omega T))|$, the digital filter response. The error, or distortion, due to spectral overlap in sampled systems, is often called aliasing, see Fig. 7.1. If the sampling frequency is sufficiently high, the aliasing can be either avoided or minimised. This applies only to narrow band lowpass or bandpass filters, but due to aliasing, highpass filters can not be designed by the impulse invariance method. Another interpretation of the aliasing is obtained from the (s,z) mapping which, as pointed out in Eq. (1.14), for this method is given by

$$z = e^{sT} \tag{7.2}$$

This relationship has been briefly examined in Section 1.3 and illustrated in Fig. 1.3. The mapping from the analogue frequency Ω to the frequency variable ω in the digital domain is many-to-one, which reflects the effects of aliasing due to sampling. The impulse invariance method is not much used, but it is useful here to motivate our interest in the bilinear transformation.

The bilinear transform, in the next section, eliminates aliasing but introduces distortion of the frequency axis.

Example 7.1

Suppose that we wish to design a lowpass digital filter based on a three-pole analogue filter which has the transfer function (see derivation of Eq. (7.8) in Section 7.4)

$$H(s) = \frac{-s_1 s_2 s_3}{(s - s_1)(s - s_2)(s - s_3)} \qquad (H(0) = 1) \quad .$$

Expanding H(s) in partial fractions, we obtain

$$H(s) = \frac{c_1}{s - s_1} + \frac{c_2}{s - s_2} + \frac{c_3}{s - s_3} \quad ,$$

where

$$c_1 = -(s_1 s_2 s_3)/(s_1 - s_2)(s_1 - s_3), \quad c_2 = -(s_1 s_2 s_3)/(s_2 - s_1)(s_2 - s_3) \quad ,$$

$$c_3 = -(s_1 s_2 s_3)/(s_3 - s_1)(s_3 - s_2) \quad .$$

The values of poles s_1, s_2, s_3 depend on the type of approximating function (Butterworth, Chebyshev, see, section 7.4). Choosing Butterworth, we have

$$s_1 = -\Omega_c, \quad s_2 = -\left[\left(1 - j\sqrt{3}\right)/2\right]\Omega_c, \quad s_3 = s_2^* \quad,$$

then the coefficients become

$$c_1 = \Omega_c, \quad c_2 = 2\Omega_c\left(-3 + j\sqrt{3}\right)^{-1}, \quad c_3 = c_2^* \quad,$$

where Ω_c is the analogue filter cutoff frequency (often taken as 1r/s, but here left as a design parameter). The impulse invariant design is obtained using Eq. (7.1), which produces

$$H(z) = \frac{m}{1 - e^{-m}z^{-1}} - \frac{m - me^{-m/2}\left(\cos a + (1/\sqrt{3})\sin a\right)z^{-1}}{1 - \left(2e^{-m/2}\cos a\right)z^{-1} + e^{-m}z^{-2}} \quad,$$

where $m = \Omega_c T$, $a = m\sqrt{3}/2$.

If we choose $\Omega_c = \omega_s/4$, where ω_s is the sampling frequency in rad/s, then $m = \pi/2$ and $a = \pi\sqrt{3}/4$. In this case H(z) becomes

$$H(z) = \frac{1.57}{1 - 0.21z^{-1}} - \frac{1.57 - 0.55z^{-1}}{1 - 0.19z^{-1} + 0.21z^{-2}} \quad,$$

which can be realised as a parallel connection of a first and second order system.

7.2 THE BILINEAR TRANSFORM (BT) METHOD

The mapping relationship in this case is given by

$$s = \frac{2}{T}\frac{z-1}{z+1} \quad, \tag{7.3}$$

which is called the bilinear transformation. To find the mapping of the frequencies, we set $s = j\Omega$ and $z = e^{j\omega T}$, and obtain

$$\Omega = \frac{2}{T}\tan\frac{\theta}{2}, \qquad \theta = \omega T \quad . \tag{7.4}$$

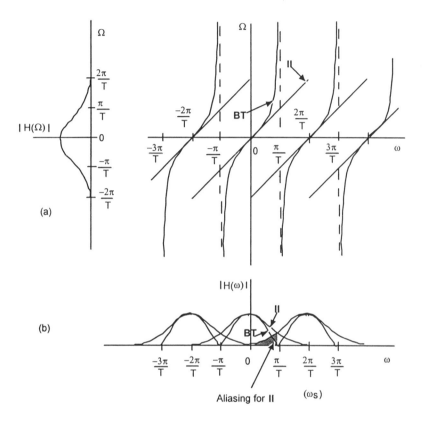

Fig 7.1:
(a) An analogue filter response
(b) Discrete time filtering response of (a) for the impulse invariance (II)
and the bilinear transform (BT) mapping

This function, plotted in Fig. 7.1, results from one-to-one mapping in which the entire range of Ω maps exactly once into the range $-\omega_s/2 < \omega < \omega_s/2$, where $\omega_s = 2\pi/T$ is the sampling frequency. Therefore, the use of the bilinear transformation eliminates the aliasing but it introduces a nonlinear frequency warping (frequency compensation) which needs to be used in the design procedure.

Usually, the design of the digital filter begins with specifications in the digital domain in terms of the frequency variable ω (or $\theta = \omega T$). These specifications are converted to the analogue domain using Eq. (7.4). The analogue filter is then designed so that it meets the specification and it is converted to a digital filter using Eq. (7.3). In this procedure, the parameter T may be set to any arbitrary value, for example T=1.

Example 7.2

Design a lowpass filter with a 3 dB cutoff at $\theta_c = 0.3\ \pi$, by using the bilinear transformation applied to the analogue filter

$$H(s) = \frac{\Omega_c}{s + \Omega_c} \quad ,$$

where Ω_c is the 3 dB cutoff of the analogue filter.

Solution:

$\theta_c = 0.3\ \pi$ corresponds to

$$\Omega_c = \frac{2}{T}\tan 0.15\pi = \frac{0.94}{T} \quad ,$$

$$H(s) = \frac{0.94/T}{s + 0.94/T} \quad .$$

Applying now the bilinear transform, Eq. (7.3), we obtain

$$H(z) = \frac{0.32\left(1 + z^{-1}\right)}{1 - 0.36z^{-1}} \quad ,$$

where the parameter T is cancelled out. The frequency response of this digital filter is

$$H(\theta) = \frac{0.32\left(1 + e^{-j\theta}\right)}{1 - 0.36e^{-j\theta}}, \qquad \theta = \omega T \ .$$

At $\omega = 0$, $H(0) = 1$, and at $\theta = 0.3\pi$, we obtain $|H(0.3\pi)| = 0.707$, as required by the specification.

Example 7.3

For the Butterworth filter in Example 7.1 the analogue filter has transfer function

$$H(s) = \frac{\Omega_c^3}{\left(s + \Omega_c\right)\left(s^2 + s\Omega_c + \Omega_c^2\right)} \quad .$$

Assume that we want the cutoff frequency for the discrete-time (or digital) filter to be $\theta_c = \omega_c T = \pi/2$. Then to compensate for the frequency warping, we must have the analogue filter cutoff frequency $\Omega_c = \tan(\theta_c/2) = \tan(\pi/4) = 1$.

Therefore,

$$H(s) = \frac{1}{(s+1)\left(s^2 + s + 1\right)}$$

and the corresponding digital filter is obtained as follows.

$$H\left(s = \frac{z-1}{z+1}\right) = H(z) = \frac{\left(1 + z^{-1}\right)^3}{2\left(3 + z^{-2}\right)}.$$

The frequency response of this filter is

$$H(\theta) = \frac{\left(1 + e^{-j\theta}\right)^3}{2\left(3 + e^{-j2\theta}\right)},$$

which needs to be calculated and plotted over the range $0 < \theta < \pi$. (Note that, in this example, the factor $2/T$ in Eqs. (7.3) and (7.4) has been left out. This simplifies the expressions and it has no effect on the result).

7.3   FREQUENCY TRANSFORMATIONS

In the case of FIR filters one can deal with any type of filter specified in the frequency domain, by performing the integration in Eq. (6.9) or using Eq. (6.16). The approach is different for IIR filters. It will be noticed that design examples for IIR filters were all for the lowpass type. To obtain other types of filter such as highpass, bandpass or bandstop, the procedure is based on the use of frequency transformation tables. There are several ways of performing these transformations. The most used one is shown in Table 7.1 , where digital transformations are given for converting a prototype digital lowpass filter into either a highpass, bandpass or bandstop or another lowpass digital filter [23,p230].

The design parameters in Table 7.1 are:

ω_p, is the cutoff frequency for the prototype digital LP filter

ω_c, is the desired cutoff frequency for another LP or HP filter

ω_1, ω_2, are the desired upper and lower cutoff frequencies for BP and BS filters

Note that we are using here ωT in place of $\theta = \omega T$.

Filter type	Transformation	Design parameters
Lowpass	$z^{-1} \to \dfrac{z^{-1} - \alpha}{1 - \alpha z^{-1}}$	$\alpha = \dfrac{\sin\left(\dfrac{\omega_p - \omega_c}{2}\right)T}{\sin\left(\dfrac{\omega_p + \omega_c}{2}\right)T}$ ω_c = desired cutoff frequency
Highpass	$z^{-1} \to \quad - \dfrac{z^{-1} + \alpha}{1 + \alpha z^{-1}}$	$\alpha = -\dfrac{\cos\left(\dfrac{\omega_p + \omega_c}{2}\right)T}{\cos\left(\dfrac{\omega_p - \omega_c}{2}\right)T}$ ω_c = desired cutoff frequency
Bandpass	$z^{-1} \to - \dfrac{z^{-2} - \dfrac{2\alpha k}{k+1}z^{-1} + \dfrac{k-1}{k+1}}{\dfrac{k-1}{k+1}z^{-2} - \dfrac{2\alpha k}{k+1}z^{-1} + 1}$	$\alpha = \cos \omega_0 T = \dfrac{\cos\left(\dfrac{\omega_2 + \omega_1}{2}\right)T}{\cos\left(\dfrac{\omega_2 - \omega_1}{2}\right)T}$
		$k = \cot\left(\dfrac{\omega_2 - \omega_1}{2}\right)T \ \tan\left(\dfrac{\omega_p T}{2}\right)$ ω_2, ω_1 = desired upper and lower cutoff frequencies
Bandstop	$z^{-1} \to \dfrac{z^{-2} - \dfrac{2\alpha}{1+k}z^{-1} + \dfrac{1-k}{1+k}}{\dfrac{1-k}{1+k}z^{-2} - \dfrac{2\alpha}{1+k}z^{-1} + 1}$	α as in the above $k = \tan\left(\dfrac{\omega_2 - \omega_1}{2}\right)T \ \tan\left(\dfrac{\omega_p T}{2}\right)$

Table 7.1

Example 7.4

Convert the single-pole lowpass filter with the transfer function

$$H(z) = \frac{0.32\left(1 + z^{-1}\right)}{1 - 0.36 z^{-1}}$$

into a bandpass filter with upper and lower cutoff frequencies $\theta_2 = \omega_2 T = 3\pi/5$ and $\theta_1 = \omega_1 T = 2\pi/5$ respectively. The lowpass filter has 3dB bandwidth $\theta_p = \omega_p T = 0.3\pi$ (as in Example 7.2).

Solution

The desired transformation is given in Table 7.1, for which we have to determine

$$\alpha = \frac{\cos(\theta_2 + \theta_1)/2}{\cos(\theta_2 - \theta_1)/2} = \frac{\cos \pi/2}{\cos \pi/10} = 0 \quad ,$$

$$k = \cot \frac{\theta_2 - \theta_1}{2} \tan \frac{\theta_p}{2} = \cot \frac{\pi}{10} \tan 0.15\pi = 3.08 \times 0.51 = 1.57 \quad ,$$

$$\frac{k-1}{k+1} = \frac{0.57}{2.57} = 0.22 \quad .$$

The transformation for this case is

$$z^{-1} \rightarrow -\frac{z^{-2} + 0.22}{0.22z^{-2} + 1} \quad ,$$

which after the substitution in the lowpass H(z) and a little algebra produces the bandpass transfer function as

$$H(z) = \frac{0.23\left(1 - z^{-2}\right)}{1 + 0.537z^{-2}} \quad .$$

More detailed designs with frequency transformations can be found, for example, in [16, Section 8.5].

7.4   DESIGN OF ANALOGUE LOWPASS FILTER PROTOTYPES

If the transfer function of an analogue filter H(s) is given and we want to implement it as an IIR filter, one can use either the impulse invariance method or, more often, the bilinear transform as discussed in Sections 7.1 and 7.2 respectively. On the other hand, if a digital filter is specified and we want to realise it in IIR form, it is again required to obtain the equivalent analogue prototype filter transfer function H(s) and transform it to the digital form H(z).

The analogue prototype filter used for this purpose is always of lowpass type. Other filter types are obtained by frequency transformations in the analogue domain ([27], Section 8.3.1) or in the digital domain using Table 7.1. A typical prototype lowpass filter specification, Fig. 7.2 is given in terms of the magnitude-frequency response with tolerance limits in the passband $(1-\delta_1$, for $0 \leq \Omega \leq \Omega_p)$, and in the stopband $(\delta_2$, for $\Omega > \Omega_s)$. The phase response is

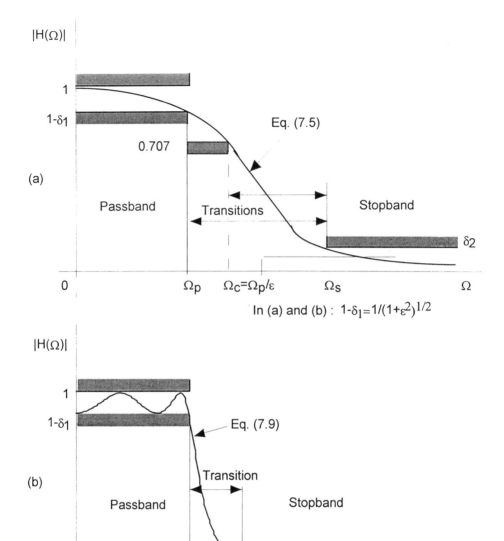

Fig 7.2: Tolerance limits for approximation of ideal lowpass filter
and frequency responses of typical approximations
(a) Butterworth and
(b) Chebychev 1

not considered. This specification can be approximated in a number of ways.

The Butterworth approximating function Fig. 7.2(a) has a smooth response in both the pass and stop bands, while Chebyshev I Fig. 7.2(b) has a controlled ripple response in the passband, but smooth response in the stopband. Other approximating functions are : Chebyshev II, with smooth response in the passband, and ripple in the stopband, and elliptic (or Cauer) function with ripple response in both pass and stopbands [16, 25, 27, 34]. We consider here only the two types indicated in Fig. 7.2.

The **lowpass Butterworth filters** have the magnitude-squared frequency response

$$|H(\Omega)|^2 = \frac{1}{1 + \left(\dfrac{\Omega}{\Omega_c}\right)^{2N}} = \frac{1}{1 + \varepsilon^2 \left(\dfrac{\Omega}{\Omega_p}\right)^{2N}} \quad , \tag{7.5}$$

where N is the order of the filter, and Ω_c is the cutoff frequency (-3 dB); Ω_p is the passband edge frequency and $1/(1+\varepsilon^2)$ is the band-edge value. In our designs, we need to determine the filter order N, which satisfies a given specification, and then its transfer function H(s).

One can show [27] that the filter order is given by

$$N = \frac{\log_{10}\left[\left(1/\delta_2^2\right) - 1\right]}{2 \log_{10}\left(\Omega_s/\Omega_c\right)} \quad , \tag{7.6}$$

where the parameters δ_2, Ω_c, Ω_s are specified in Fig. 7.2 (a). The corresponding H(s) can then be found from tables giving poles in the s-plane or giving the polynomial. It is not always easy to find such tables, but we can determine poles from the following equation

$$s_k = \Omega_c \exp\left\{j\left[\frac{\pi}{2} + \frac{(2k+1)\pi}{2N}\right]\right\}, \quad k = 0,1,\ldots,N-1 \quad . \tag{7.7}$$

For N = 3, we have

$$s_0 = e^{j2\pi/3}\Omega_c$$
$$s_1 = e^{j\pi}\Omega_c = -\Omega_c$$
$$s_2 = e^{j4\pi/3}\Omega_c = e^{-j2\pi/3}\Omega_c$$

and the transfer function is given by

$$H(s) = (-1)^N \frac{s_0 s_1 s_2}{(s - s_0)(s - s_1)(s - s_2)}$$

$$= \frac{\Omega_c^3}{(s + \Omega_c)\left[s^2 - \left(2\cos\frac{2\pi}{3}\right)\Omega_c\, s + \Omega_c^2 \right]}$$

$$\therefore \qquad H(s) = \frac{\Omega_c^3}{(s + \Omega_c)(s^2 + s\Omega_c + \Omega_c^2)} \qquad . \tag{7.8}$$

This is the transfer function used in Example 7.3. In analogue filter tables $\Omega_c = 1$. When $\Omega_c \neq 1$ is reintroduced, the procedure is to divide s by the new $\Omega_c \neq 1$.

The **lowpass Chebyshev type I** Fig. 7.2(b) filters have the magnitude-squared frequency response given as

$$|H(\Omega)|^2 = \frac{1}{1 + \varepsilon^2\, C_N^2\left(\dfrac{\Omega}{\Omega_p}\right)} \qquad , \tag{7.9}$$

where ε is a parameter of the filter related to the ripple in the passband (see Fig.7.2), and $C_N(x)$ is the Nth-order Chebyshev polynomial [16, 27].

As in the Butterworth case, we consider determination of N and H(s). For a given specification $\varepsilon, \delta_2, \Omega_s/\Omega_p$ we have

$$N = \frac{\cosh^{-1}\left(1/\varepsilon\delta_2\right)}{\cosh^{-1}\left(\Omega_s/\Omega_c\right)} \qquad , \tag{7.10}$$

where $\varepsilon^2 = \left[1/(1 - \delta_1)^2 \right] - 1 \qquad .$

The procedure for determining poles is not as simple as for the Butterworth case. The poles lie on an ellipse in the s-plane with major (r_1) axis, and minor (r_2) axis, given by

$$r_1 = \frac{\beta^2 + 1}{2\beta} \Omega_p, \quad r_2 = \frac{\beta^2 - 1}{2\beta} \Omega_p \quad ,$$

$$\text{where} \quad \beta = \left[\frac{\sqrt{1 + \varepsilon^2} + 1}{\varepsilon} \right]^{1/N} \quad .$$

The poles are then obtained as follows:

$$s_k = (r_2 \cos \phi_k) + j(r_1 \sin \phi_k) \quad , \tag{7.11}$$

$$\text{where} \quad \phi_k = \frac{\pi}{2} + \frac{(2k + 1)\pi}{2N} \qquad k = 0, 1, 2, \ldots, N - 1 \quad .$$

With poles determined, we can form the transfer function in the same way as shown for the Butterworth 3rd order filter. For example, a Chebyshev I filter with 1 dB ripple and $N = 3$, has the following transfer function

$$H(s) = \frac{0.49\,14\,\Omega_p^3}{\left(s + 0.4942\,\Omega_p\right)\left[s^2 + \left(0.4942\,\Omega_p\right)s + 0.9943\,\Omega_p^2\right]} \quad . \tag{7.12}$$

7.5   IIR STRUCTURES AND IMPLEMENTATIONS

We have seen, in Sections 7.1 to 7.3, that IIR transfer functions are of general type as in Eq. (1.25). The difference equation corresponding to Eq. (1.25) is given by Eq. (1.24), and the IIR structure is given as a cascade of two units, Fig. 1.4, known as direct form I.

By interchanging the order of these two units, the overall system response remains the same (for linear time-invariant systems). So, by reversing the order of the units in Fig. 1.4 we obtain the structure, Fig. 7.3(a).

This structure is the cascade of

$$w(n) = x(n) - \sum_{m=1}^{N} a_m w(n - m) \tag{7.13}$$

and

$$y(n) = \sum_{m=0}^{N} b_m w(n - m) \quad . \tag{7.14}$$

The two branches with transmission z^{-1} have the same inputs, hence only one branch is required. Therefore, this configuration can be redrawn as in Fig.

7.3(b) requiring N delays and 2N+1 multiplications. Since it requires the minimum number of delays, this direct form II is sometime called a canonic form.

　　The direct forms I and II can be employed to implement digital filters, both in hardware and in software. However, this can cause two major problems: firstly lack of hardware flexibility and secondly the sensitivity to changes in coefficients increases with the order of the filter. These can be avoided by using cascades or parallel connections of first and second order sections. (Appendix 7.1).

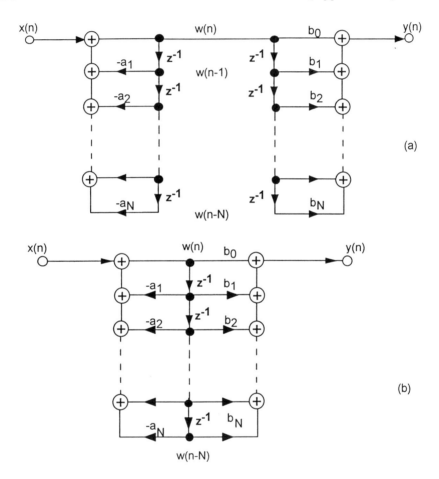

Fig. 7.3:
　　(a) Interchange of units in Fig. 1.4
　　(b) Reduction of (a) into direct form II (canonic form)

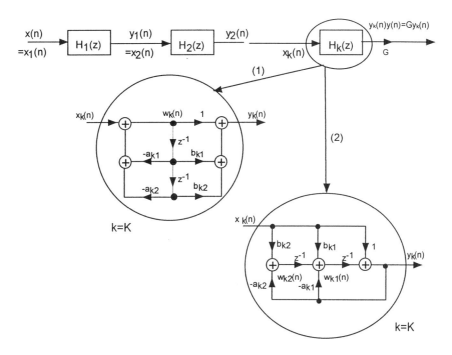

Fig. 7.4: Cascade connection
(1) Direct form (canonic-form) block
(2) Transposed form block

For the **cascade connection**, Fig. 7.4, the transfer function is

$$H(z) = G\prod_{k=1}^{K} H_k(z) \quad ,$$

$$(7.15)$$

where G is the fixed gain parameter, and $H_k(z)$ has the general form

$$H_k(z) = \frac{1 + b_{k1}z^{-1} + b_{k2}z^{-2}}{1 + a_{k1}z^{-1} + a_{k2}z^{-2}} \quad .$$

$$(7.16)$$

The coefficients $\{a_{ki}\}$ and $\{b_{ki}\}$ are real. To preserve the modularity in the implementation of H(z), it is often preferable to use the basic second-order blocks, in direct-form or transpose-form, Fig.7.4.

The computational algorithm for the **direct-form**, Fig. 7.4(1) is given by the following set of equations.

$$y_0(n) = x(n)$$
$$w_k(n) = y_{k-1}(n) - a_{k1} w_k(n-1) - a_{k2} w_k(n-2)$$
$$y_k(n) = w_k(n) + b_{k1} w_k(n-1) + b_{k2} w_k(n-2) \ ,$$

where, $k = 1,2,...,K$, and

$$y(n) = G \ y_K(n) \ . \tag{7.17.1}$$

Similarly, the algorithm for the **transpose-form**, Fig. 7.4(2), is

$$y_0(n) = x(n)$$
$$y_k(n) = y_{k-1}(n) + w_{k1}(n-1)$$
$$w_{k1}(n) = b_{k1}y_{k-1}(n) + w_{k2}(n-1) - a_{k1} y_k(n)$$
$$w_{k2}(n) = b_{k2} y_{k-1}(n) - a_{k2} y_k(n) \ ,$$

where, $k = 1,2,...,K$, and

$$y(n) = G \ y_K(n) \ . \tag{7.17.2}$$

It can be shown easily that both these forms produce the transfer function $H_k(z)$, given by Eq. (7.16).

For the **parallel connection**, Fig. 7.5, the transfer function is

$$H(z) = C + \sum_{k=1}^{K} H_k(z) \ , \tag{7.18}$$

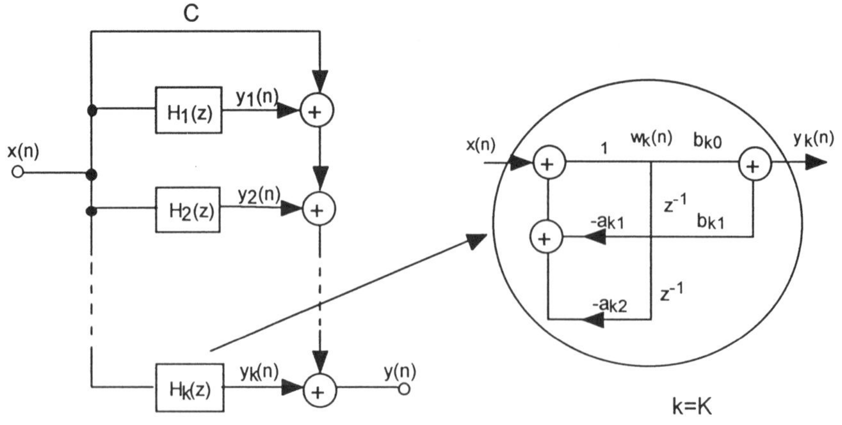

Fig 7.5: Parallel connection and its 2nd order block

where C is a constant, and $H_k(z)$ are given by

$$H_k(z) = \frac{b_{k0} + b_{k1}z^{-1}}{1 + a_{k1}z^{-1} + a_{k2}z^{-2}} \quad . \tag{7.19}$$

As in the previous cases, the coefficients are real.

The computational algorithm in this case is as follows:

$$w_k(n) = x(n) - a_{k1} w_k(n-1) - a_{k2} w_k(n-2)$$
$$y_k(n) = b_{k0} w_k(n) + b_{k1} w_k(n-1) \quad , \tag{7.20}$$

where $k = 1,2,\ldots,$ K and

$$y(n) = Cx(n) + \sum_{k=1}^{K} y_k(n) \quad .$$

Example 7.5

The transfer function of a system is given as

$$H(z) = 10 \frac{1 + 0.8333z^{-1} - 2z^{-2} + 0.6667z^{-3}}{1 - 1.875z^{-1} + 1.4688z^{-2} - 0.5313z^{-3} + 0.0469z^{-4}} \quad .$$

Determine the cascade and parallel realisations.

Solution:.

For this purpose it is necessary first to find roots of the numerator and denominator polynomials which is not an easy task. (roots of polynomials of order higher than second can be found by methods of algebra, or by some approximate numerical procedure using digital computers).

In the above case, the roots were available which enable writing H(z) as follows:

$$H(z) = 10 \frac{\left(1 - (0.5)z^{-1}\right)\left(1 - (2/3)z^{-1}\right)\left(1 + 2z^{-1}\right)}{\left(1 - (3/4)z^{-1}\right)\left(1 - (1/8)z^{-1}\right)\left[1 - (0.5 + j0.5)z^{-1}\right]\left[1 - (0.5 - j0.5)z^{-1}\right]} \quad .$$

The cascade realisation is easily obtained from the above by forming second order polynomials. It is clear in the denominator that to obtain real coefficients one must combine the conjugate complex pair, and then the other pair. In the numerator one can choose pairing in three ways, but to simplify we choose the first and third to form a second order polynomial.

Therefore, one possible realisation by such pairing is

$$H_1(z) = \frac{1 - (2/3)z^{-1}}{1 - (7/8)z^{-1} + (3/32)z^{-2}} \quad ,$$

$$H_2(z) = \frac{1 + (3/2)z^{-1} - z^{-2}}{1 - z^{-1} + (1/2)z^{-2}}$$

and hence

$$H(z) = 10 \, H_1(z) \, H_2(z) \quad .$$

The coefficients in our second order sections, Eq. (7.16) and Fig. 7.4 are

$$a_{11} = -(7/8), \ a_{12} = 3/32; \ b_{11} = -(2/3), \ b_{12} = 0$$

and

$$a_{21} = -1, \ a_{22} = 1/2; \ b_{21} = 3/2, \ b_{22} = -1.$$

For the parallel realisation the procedure is not so straightforward. We need first to expand $H(z)$ into partial fractions:

$$H(z) = \frac{A_1}{1 - 0.75z^{-1}} + \frac{A_2}{1 - 0.125z^{-1}} + \frac{A_3}{1 - 0.5(1+j)z^{-1}} + \frac{A_4}{1 - 0.5(1-j)z^{-1}} \quad ,$$

where A_1 to A_4 have to be determined. This requires some arithmetic work resulting in

$$A_1 = 2.93, \ A_2 = -17.68, \ A_3 = 12.25 - j14.57, \ A_4 = A_3^* \quad .$$

Recombining pairs of poles, we have

$$H(z) = \frac{-14.75 - 12.9z^{-1}}{1 - (7/8)z^{-1} + (3/32)z^{-2}} + \frac{24.50 + 26.82z^{-1}}{1 - z^{-1} + 0.5z^{-2}} \quad .$$

The coefficients for two sections, Eq. (7.19) and Fig. 7.5 are

$$a_{11} = -(7/8), \ a_{12} = 3/32, \ b_{10} = -14.75, \ b_{11} = -12.9$$

and $\quad a_{21} = -1, \ a_{22} = 1/2, \ b_{20} = 24.50, \ b_{21} = 26.82.$

7.6 QUANTIZATION EFFECTS

Most discrete-time systems are realised by digital computers or digital hardware, hence the amplitudes of signals and coefficients are quantized discrete variables. We assume the quantized quantities to be represented as 2's complement fixed point fractions of length (b+1) bits (one to the left and b to the right of the binary point).

Filter coefficients are quantized once, in the design process, and their values remain constant in the filter implementation. The effect of coefficient quantization is to perturb h(n), H(z) from their ideal, designed, values in a deterministic manner. We can check the quantized design, and if necessary redesign it, and/or allocate more bits to satisfy the specification.

Signal x(n) quantization, due to rounding or truncation occurs at the input (ADC noise, Section 1.2) and internally with the multipliers during processing (filtering). It is best viewed as a random process.

It is especially important to consider quantization effects for implementations in fixed-point arithmetic, such as with microprocessors or special digital hardware. Larger computers using floating-point arithmetic of 32 bits or more are usually sufficiently accurate so that these effects can be neglected. Signal quantization, for example, is closely related to scaling and dynamic range considerations. Self-scaling of floating-point arithmetic largely eliminates quantization problems. Therefore, our study of quantization effects refers to the fixed-point case.

7.6.1 Quantization noise from filters

The basic arithmetic operations in digital filter implementations are multiplication and addition. For the fixed-point arithmetic, the result of multiplication must be rounded or truncated. We consider the effects of rounding in IIR type filters, but results can be easily modified for truncation, [27].

For fixed-point arithmetic the product of two b-bits numbers results in numbers 2b-bits long, which need to be rounded to b-bits. Therefore, each multiplication between the signal and fixed coefficient, after rounding will produce roundoff noise similar to ADC noise. For example, the product $a_i w(n-i)$ resulting in 2(b+1) bits, after rounding, is represented by $a_i w(n-i)$ with (b+1) bits and the error $e_i(n)$ added to the sum node for each multiplier. These are shown, in Fig.7.6, as e_1, e_2 at the input node and e_3, e_4 at the output node. At the input node we have the signal x(n), and its quantization noise e (Section 1.2), and the product quantization errors e_1 and e_2. Since these signals are low level and independent we can express the variance of the output noise due to e as follows.

$$\sigma_f^2 = \sigma_e^2 \sum_{n=-\infty}^{\infty} h^2(n) = \sigma_e^2 \left[\frac{1}{2\pi j} \oint H(z) H(z^{-1}) z^{-1} \, dz \right] , \qquad (7.21)$$

where h(n) is the impulse (or unit sample) response from the input node to the output node, and H(z) is the corresponding transfer function. The above expression can be easily computed by means of the contour integration which enables us to express the quantity between square brackets as

$$\sum \text{Residue} \left[H(z) H\left(z^{-1}\right) z^{-1} \right] \text{ at poles inside unit circle}$$

$$= \sum_i \left\{ (z - p_i) \left[H(z) H\left(z^{-1}\right) z^{-1} \right] \right\}_{z = p_i} \tag{7.22}$$

p_i are poles within the unit circle in the z-plane.

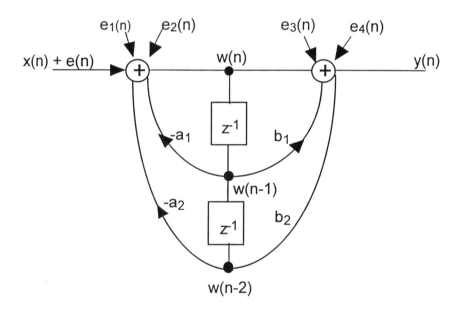

Fig. 7.6: Product quantization noise model for the direct (canonic) 2nd order section

The output variance, in Fig. 7.6, for three quantization errors at the input node and two at the output node is given by

$$\sigma_f^2 = 3\sigma_e^2 \sum_{n=-\infty}^{\infty} h^2(n) + 2\sigma_e^2 \quad , \tag{7.23}$$

where σ_e^2 is the variance of the individual noise sources with the value of $\sigma_e^2 = q^2/12$ (as in Section 1.2)

Example:

To illustrate the calculation we assume a network in Fig. 7.6 with the following transfer function

$$H(z) = \frac{1 - 0.8\,z^{-1} + 0.12\,z^{-2}}{1 - 1.2\,z^{-1} + 0.32\,z^{-2}} \quad,$$

which means $a_1 = -1.2$, $a_2 = 0.32$, $b_1 = -0.8$, $b_2 = 0.12$. In this case, it is easy to find the poles and zeros and we can rewrite H(z) as

$$H(z) = \frac{(z - 0.2)(z - 0.6)}{(z - 0.4)(z - 0.8)} \quad.$$

The output quantization noise variance, for this case is given by Eq. (7.23). The main calculation, indicated earlier in Eqs. (7.21) and (7.22), is given by

$$\sum_{n=-\infty}^{\infty} h^2(n) = \frac{1}{2\pi j} \oint H(z)\,H(z^{-1})\,z^{-1}\ dz \quad,$$

$$= \sum_i \left\{ (z - p_i)\left[H(z)H(z^{-1})z^{-1} \right] \right\}_{z = p_i}$$

where p_i are poles within the unit circle in the z-plane. For H(z) given in this example, we have

$$\sum_{n=-\infty}^{\infty} h^2(n) = \left\{ (z - 0.4)\frac{(z - 0.2)(z - 0.6)}{(z - 0.4)(z - 0.8)}\frac{(1 - 0.2z)(1 - 0.6z)}{(1 - 0.4z)(1 - 0.8z)}z^{-1} \right\}_{z=0.4}$$

$$+ \left\{ (z - 0.8)\frac{(z - 0.2)(z - 0.6)}{(z - 0.4)(z - 0.8)}\frac{(1 - 0.2z)(1 - 0.6z)}{(1 - 0.4z)(1 - 0.8z)}z^{-1} \right\}_{z=0.8}$$

$$= 0.122 \times 2.5 + 0.535 \times 1.25 = 0.974$$

Using this result in Eq. (7.23) we have

$$\sigma_f^2 = \sigma_e^2 (3 \sum_{n=-\infty}^{\infty} h^2(n) + 2)$$

$$\sigma_f^2 = 4.92\sigma_e^2$$

We have considered a second-order section and shown how the quantization noise produced by the multipliers is handled. In practice, filters are of higher order and, as pointed out earlier, realised in cascade or parallel forms. For the cascaded form the analysis is more complicated but a general expression can be formulated to enable a systematic analysis [11]. On the other hand the analysis of the second-order filter can be applied directly to higher-order filters based on a parallel realisation. In this case, each second-order section is independent of all the other sections, and hence the total quantization noise power is simply the linear sum of the quantization noise powers of each of the individual sections.

In the case of FIR filters, the output noise due to rounding the products is independent of the filter parameters. This noise is not processed by the filter and appears directly at the output as $N\sigma_e^2$ where N is the length of the FIR filter response [23].

7.6.2   Scaling to prevent overflow

The calculation of roundoff noise presented in the previous section is meaningless if the filter is not correctly scaled. The roundoff noise is the dominant component of output noise in a digital filter only when overflows in internal registers are negligible. Internal overflows cause large errors and must be prevented. This is accomplished by properly scaling the realisation. Scaling constrains the numerical values of internal filter variables to remain in a range appropriate   to the hardware.

No overflow condition at k-th node of the system can be expressed as

$$\left| y_k(n) \right| = \left| \sum_{m=-\infty}^{\infty} h_k(n)x(n-m) \right| \le \sum_{m=-\infty}^{\infty} \left| h_k(n) \right| \left| x(n-m) \right| \ , \tag{7.24}$$

where $y_k(n)$ is the system response at the k-th node when the input sequence is x(n), and the unit sample response between the node and the input is $h_k(n)$. All quantities $\left| y_k(n) \right|, \left| h_k(n) \right|$ etc. are indicated as absolute values. If x(n) is upper bounded by C_x, then

$$\left| y_k(n) \right| \le C_x \sum_{m=-\infty}^{\infty} \left| h_k(m) \right|$$

and if the dynamic range of the computer is limited to (-1,1), the condition

$$\left| y_k(n) \right| < 1$$

will be satisfied by scaling the input so that

$$C_x < \frac{1}{\displaystyle\sum_{m=-\infty}^{\infty} \left| h_k(m) \right|} \tag{7.25}$$

for all possible nodes in the system.

For an FIR filter the limits are m = 0 to M - 1, so that the sum contains M non-zero terms. The method of scaling discussed above is not always easy to apply. An alternative method based on the frequency response seems to be more attractive [11, p.443].

A stable digital filter implemented with infinite precision arithmetic has a unit sample response decaying asymptotically toward zero. For the same filter, implemented with finite register length arithmetic, the output may decay into a non-zero or oscillatory response. This effect is referred to as zero-input limit cycle behaviour, and it is a consequence of the non-linear quantizers in the feedback loop of the recursive (IIR type) filter. The limit cycle is complex and difficult to analyse, but useful results can be obtained by analysis of the first and second order systems [27].

A more serious type of limit cycle can occur due to summation noise overflow in IIR filters. This limit cycle results in large periodic oscillations, which can be avoided by proper scaling of the input data as discussed in the previous section or by using a saturating adder [27].

7.7   IIR FILTER DESIGN

A prototype lowpass (plp) filter is specified in Fig. 7.7. The desired filter characteristics are specified for magnitude response only, but the phase response (not under the designer's control) is accepted as it comes.

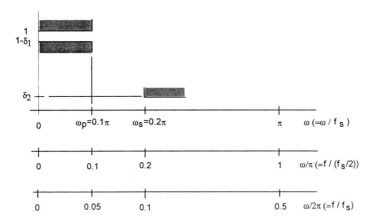

Passband ripple: $20 \log(1-\delta_1) = -1 \text{dB} : (1 - \delta_1) = 0.8913$
Stopband ripple : $20 \log \delta_2 = -40 \text{ dB} : \delta 2 = 0.01$
Sampling frequency : fs = 1 Hz

Fig. 7.7: Prototype lowpass (PLP) filter specification

Three different frequency scales are shown since they may be met in digital filter designs. For example, the second scale (ω/π) is used in Matlab programs. Also, note that these scales are normalised to the sampling frequency (f_s) which is often taken as $f_s = 1$ Hz.

The plp specification in Fig. 7.7 has been chosen arbitrarily as convenient to illustrate the use of Table 7.1. In order to meet the specification for the digital filter in Fig. 7.7, we have to convert it first to the analogue domain and then determine the analogue transfer function using one of the approximating functions in Section 7.4. Therefore, we first find the analogue, or so-called pre-warped, frequencies using Eq. (7.4) and obtain

$$\Omega_p = \tan \omega_p/2 = \tan (0.05\pi) = 0.158 \quad ,$$
$$\Omega_s = \tan \omega_s/2 = \tan (0.125\pi) = 0.414 \quad .$$

(To simplify, the factor (2/T) has been left out; see the comment at the end of Example 7.3).
We choose in this case Chebyshev 1 approximating function, Fig. 7.2(b) and Eq. (7.9), and for our plp we obtain $\varepsilon^2 = 0.259$, $N = 4$ (calculated value 3.7)

To determine H(s), for $N = 4$, we proceed as follows. Using equations in section 7.4, we obtain

$\beta = 1.43$; $r_1 = 1.065 \, \Omega_p$, $r_2 = 0.365 \, \Omega_p$.
$\phi_k = (\pi/2) + \pi(2k + 1)/8$, for $k = 0, 1, 2, 3$ produces
$\phi_0 = 5\pi/8$; $\phi_1 = 7\pi/8$; $\phi_2 = 9\pi/8$; $\phi_3 = 11\pi/8$.

The pole positions are now obtained, from Eq. (7.11), as

$$\left.\begin{array}{rcl} s_{0,3} &=& \left(-0.14 \pm j\,0.984\right)\Omega_p \\ s_{1,2} &=& \left(-0.337 \pm j\,0.408\right)\Omega_p \end{array}\right\} \qquad (7.26)$$

and the transfer function is now formed as follows

$$H(s) = \frac{1}{\sqrt{1+\varepsilon^2}} \frac{s_0 \, s_1 \, s_2 \, s_3}{(s-s_0)(s-s_3)(s-s_1)(s-s_2)} \quad .$$

For $N = 4$, $|H(0)| = 1/\sqrt{1+\varepsilon^2}$, but for N odd $|H(0)| = 1$ [27, p. 636].
Substituting for s_0 to s_3 we obtain

$$H(s) = \frac{0.247\ \Omega_p^4}{\left(s^2 + 0.28s\ \Omega_p + 0.988\ \Omega_p^2\right)\left(s^2 + 0.674s\ \Omega_p + 0.28\ \Omega_p^2\right)} \qquad (7.27)$$

In our case $\Omega_p = 0.158$, so that the above becomes

$$H(s) = \frac{0.247}{\left(40s^2 + 1.772s + 0.988\right)\left(40s^2 + 4.2658s + 0.28\right)} \qquad (7.28)$$

(In text books on analogue filter design H(s) is usually normalised to $\Omega_p = 1$ r/s. To denormalise it, we use the substitution $s \rightarrow s/\Omega_p$, where Ω_p in our case is 0.158.)

Applying now the BT substitution $s = (z - 1)/(z + 1)$, to (7.28), we obtain the corresponding discrete-time filter plp transfer function arranged in the following way

$$H(z) = \left(1.2 \times 10^{-2}\ \frac{1 + 2z^{-1} + z^{-2}}{1 - 1.8247\,z^{-1} + 0.917\,z^{-2}}\right)$$

$$\times\ \left(1.2 \times 10^{-2}\ \frac{1 + 2z^{-1} + z^{-2}}{1 - 1.7833\,z^{-1} + 0.8055\,z^{-2}}\right) \qquad (7.29)$$

7.7.1   Derivation of a lowpass filter from the plp

Assume we require a lowpass filter with $\omega_c = 0.2\pi$, Fig. 7.8(a). The transformation from the plp to lowpass, Table 7.1, is given by

$$z^{-1} \rightarrow \frac{z^{-1} - \alpha}{1 - \alpha\,z^{-1}} \qquad , \qquad (7.30)$$

where

$$\alpha = \frac{\sin(\omega_p - \omega_c)T/2}{\sin(\omega_p + \omega_c)T/2} = -0.3446$$

with $\omega_p = 0.1\ \pi$, $\omega_c = 0.2\pi$, and $T = 1$ (ie $f_s = 1$Hz).

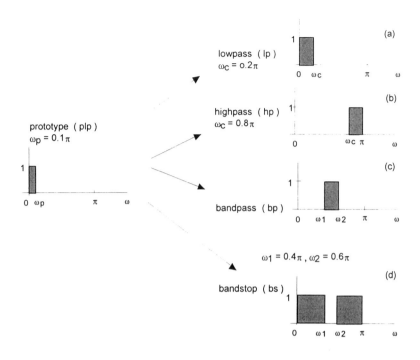

Fig. 7.8: Prototype lowpass (PLP) transformations to other filters given in Table 7.1

Using the transformation (7.30), with $\alpha = -0.3446$, in Eq. (7.29) we obtain the lowpass filter transfer function as

$$H(z) = \left(4.369 \times 10^{-2} \; \frac{1 + 2z^{-1} + z^{-2}}{1 - 1.5004\,z^{-1} + 0.8477\,z^{-2}} \right)$$

$$\times \; \left(4.2053 \times 10^{-2} \; \frac{1 + 2z^{-1} + z^{-2}}{1 - 1.5547\,z^{-1} + 0.6494\,z^{-2}} \right)$$

(7.31)

The transformation given by (7.30) can be used to check the plp and lowpass filter characteristic frequencies. For this purpose we can write (7.30) as

$$Z^{-1} = \frac{z^{-1} - \alpha}{1 - \alpha z^{-1}} \; ,$$

where Z stands for the plp and z for the lowpass.

Solving this equation for z^{-1} we obtain

$$z^{-1} = \frac{Z^{-1} + \alpha}{1 + \alpha Z^{-1}} \quad .$$

Introducing now $z = e^{j\omega}$ and $Z = e^{j\omega'}$ we can relate ω' (plp) to ω (lowpass) frequencies.

7.7.2   Derivation of a highpass filter from the plp

The specification is shown in Fig. 7.8(b). From Table 7.1, the transformation is given by

$$z^{-1} \rightarrow \frac{z^{-1} - 0.3445}{1 - 0.3445\, z^{-1}} \quad , \tag{7.32}$$

since

$$\alpha = -\frac{\cos\left(\omega_p + \omega_c\right)T/2}{\cos\left(\omega_p - \omega_c\right)T/2} = -0.3445 \quad ,$$

with $\omega_p = 0.1\pi$, $\omega_c = 0.8\pi$ and $T = 1$. Using the transformation (7.32) in (7.29) we obtain the highpass transfer function as

$$H(z) = \left(4.3658 \times 10^{-2}\ \frac{1 - 2z^{-1} + z^{-2}}{1 + 1.499\, z^{-1} + 0.8478\, z^{-2}} \right)$$

$$\times \left(4.2031 \times 10^{-2}\ \frac{1 - 2z^{-1} + z^{-2}}{1 + 1.5547\, z^{-1} + 0.6496\, z^{-2}} \right) \quad . \tag{7.33}$$

7.7.3   Derivation of a bandpass filter from the plp

The specification is shown in Fig. 7.8(c). From Table 7.1, the transformation is given by

$$z^{-1} \rightarrow \frac{0.3445 - z^{-2}}{1 - 0.3445\, z^{-2}} \quad , \tag{7.34}$$

since $\alpha = 0$ and $k = 0.487$ for $\omega_p = 0.1\pi$, $\omega_1 = 0.4\pi$, $\omega_2 = 0.6\pi$, $T = 1$. Substituting (7.34) into (7.29) one can derive the following bandpass transfer function

$$H(z) = \left(4.367 \times 10^{-2} \, \frac{1 - 2z^{-2} + z^{-4}}{1 + 1.5004\,z^{-2} + 0.8478\,z^{-4}} \right)$$

$$\times \left(4.20 \times 10^{-2} \, \frac{1 - 2z^{-2} + z^{-4}}{1 + 1.5537\,z^{-2} + 0.6498\,z^{-4}} \right) \quad .$$

(7.35)

7.7.4   Derivation of a bandstop filter from the plp

The specification is shown in Fig. 7.8(d). The transformation from Table 7.1 is given by

$$z^{-1} \to \frac{z^{-2} + 0.9021}{0.9021z^{-2} + 1} \quad , \tag{7.36}$$

since $\alpha = 0$, $k = 0.05147$ for $\omega_p = 0.1\pi$, $\omega_1 = 0.4\pi$, $\omega_2 = 0.6\pi$, $T = 1$.

Substituting (7.36) into (7.29) one can show that the bandpass transfer function is given by

$$H(z) = \left(0.4189 \, \frac{1 + 2z^{-2} + z^{-4}}{1 + 1.4875\,z^{-2} + 0.8453\,z^{-4}} \right)$$

$$\times \left(0.8236 \, \frac{1 + 2z^{-2} + z^{-4}}{1 + 0.5772\,z^{-2} + 0.2764\,z^{-4}} \right) \quad .$$

(7.37)

7.7.5   Matlab designs

The transfer functions derived in this section can be checked with Matlab using the function

$$[b, a] = \text{cheby 1 } (n, R_p, wn, \text{'type'}) \quad ,$$

where b and a represent the numerator and denominator consisting each of $(n + 1)$ coefficients. The command inputs are: n = filter order, R_p = passband ripple in dB, wn = cutoff frequency. For all the filters in this design : $n = 4$, $R_p = 1$ dB.

Cutoffs wn for plp, lp, hp, bp, bs are respectively 0.1, 0.2, 0.8, [0.4 0.6], [0.4 0.6]. The filter 'type' needs to be specified for hp and bs as 'high' and 'stop' respectively; for lp and bp 'type' is not to be specified. For the frequency plot , the additional program, for all cases, is :

```
[ h, w ] = freq z ( b, a, 512 );
plot (w * 0.5/pi, abs (h)); grid
```

Function freqz returns both h, the complex frequency response, and w, a vector containing the n frequency points equally spaced around the upper half of the unit circle, so w contains n points between 0 and π. In plot

$$(w * 0.5/pi, abs(h));$$

the frequency scale is changed from $0, \pi$ to $0, 0.5$.

Additional program for the impulse response and step response is as follows:

```
n = 0 : 60;
x = [ 1   zeros (1 : 60 ) ];          % unit impulse input
v = [ 0   60   - 0.1   0.3 ];  axis (v);
y = filter ( b, a, x ); stem (n, y); grid; pause; clf
x = [ 1   ones (1 : 60) ];           % unit step response
v = [ 0 60 0.1 2 ] ; axis (v)
y = filter (b, a, x); stem (n, y); grid; pause; clf.
```

The length of the input sequence and axis have been chosen for the lowpass designed in this section. Results for the frequency response and impulse response are shown respectively in Figs. 7.9 and 7.10.

7.8   IIR FILTER IMPLEMENTATION

Compared to FIR filters, the implementation of IIR filters, is quite problematical on integer DSPs. For the same performance as a FIR filter, less arithmetic is normally required. However, the dynamic range of the internal variables at the nodes is often embarrassingly large, requiring the programmer to pay much more attention to scaling between sections. We shall firstly look at the performance of a practical two section IIR filter, using the transposed structure and then see how the dynamic range can be improved by scaling, using the efficient shifting mechanisms of the DSP.

The TMS320C25 IIR filter programs whose performance is described here use the LTD and MPY instructions to perform the basic multiply-accumulate, being faster than the special MACD instruction described in 3.3.2. This is faster only where a small (<8) number of iterations is required.

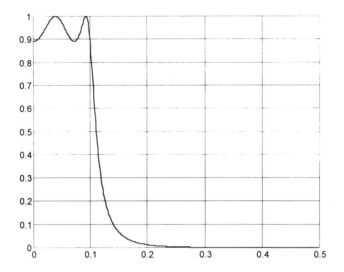

Fig. 7.9: The magnitude response of the lowpass filter using the Matlab program given in section
7.7.5

(Note that the frequency scale is 0 to 0.5, obtained by division of the scale 0 to π by 2π.
Therefore, the passband edge frequency $\omega_p = 0.2\pi$ becomes 0.1)

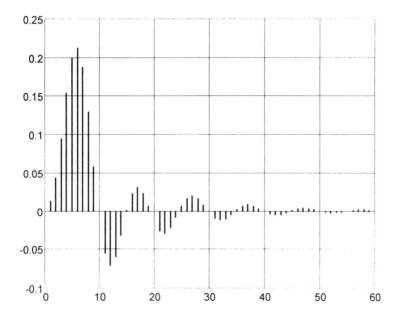

Fig. 7.10: The unit pulse response for the lowpass filter as in Fig. 7.9

7.8.1 Low-pass filter implementation

The transposed structure of the two section low-pass filter of Fig 7.4(2) was implemented on the TMS320C25 and its performance measured on the simulator. Values were scaled between +1 and -1. The algorithm, from Eq. (7.17.2) for the two section filter was implemented in 6 steps ,as shown in Fig 7.11. The coefficients calculated in Section 7.7, Eq. (7.31) are all less than 2, but not within the convenient range of $\pm$ 1. They have therefore been divided by 2. Negative signs have been incorporated in the constants, so that the program contains no subtractions, see Fig 7.13. Note that this is an algorithm, not an equation; for example the w11 in step 1 of Fig. 7.11 is that calculated for the previous sample.

```
y1 = x0 + w11
w11 = 2 * (b11/2 * x0 +  (-a11/2) * y1 + w12/2)
w12 = 2 * (b12/2  * x0 +  (-a12/2)) * y1


y2 = y1 + w21
w21 = 2 * (b21/2 * y1 +  (-a21/2) * y2 + w22/2)
w22 = 2 * (b22/2 * y1 +  (-a22/2) * y2)
```

Fig. 7.11: Implementation steps for the IIR filter

```
loop         ;start first section
             in l.x0,input
             zalh l.x0
             addh l.w11              ; acch = x0+w11

             sach l.y1,noshift       ;  -> y1
             lt l.x0
             mpy l.b11
             pac                     ; acch = x0*b11/2
             lt l.y1
             mpy l.a11
             apac                    ; + y1*a11/2
             add l.w12,shift15       ; + w12/2
             sach l.w11,shift1       ; * 2 -> w11
             lt l.x0
             mpy l.b12
             pac                     ; acch=x0*b12/2
             lt l.y1
             mpy l.a12
             apac                    ; + y1*a12/2
             sach l.w12,shift1       ; *2 -> w12
```

Fig. 7.12: TMS320C25 listing of the 1st section of the IIR filter

The TMS320C25 listing for the calculation of the first section of the IIR low pass filter is given in fig 7.12. A complete listing is given in Appendix 7.2. The 'spm' instruction has been used to set a left shift of 1 bit when products in the P register are transferred to the accumulator, so that the accumulator high part contains the correctly adjusted value appropriate to ±1 scaling.

```
;
;
;define filter constants in range  -1 to +1, transferred to data memory
coefstart
        data  24583 ; 0.7502 = -a11/2  -> data memory address a11
        data -13889 ; -0.4239 = -a12/2  -> data memory address a12
        data  25471 ; 0.7773 = -a21/2  -> data memory address a21
        data  -10640  ; 0.3247 = -a22/2 -> data memory address a22
        data  32767 ; 1.0 = b11/2      -> data memory address b11
        data  16384 ; 0.5 = b12/2      -> data memory address b12
        data  32767 ; 1.0 = b21/2      -> data memory address b21
        data  16384 ; 0.5 = b22/2      -> data memory address b22
coeflen equ $ - coefstart
```

Fig. 7.13: IIR LP filter coefficients as defined in the TMS320C25 program

Note that additions only were used, and the negative signs are incorporated in the definition of the table of coefficients, as shown in fig 7.13.

7.8.2   Performance of the TMS320C25 low pass IIR filter

The impulse response of the LP IIR filter is shown in Fig. 7.14 and is in good agreement with the 'perfect' response given in Fig. 7.10.

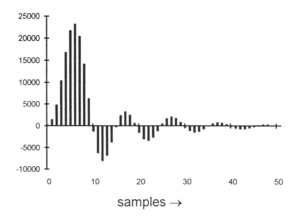

Fig. 7.14: Impulse response IIR low pass filter without internal scaling,
impulse amplitude = 200 (0,0)

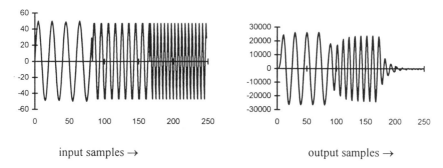

input samples →   output samples →

Fig. 7.15: Transpose form IIR LP filter,
unscaled, sine input (± 50), f/fs = 0.05, 0.1, 0.2; graph shows output

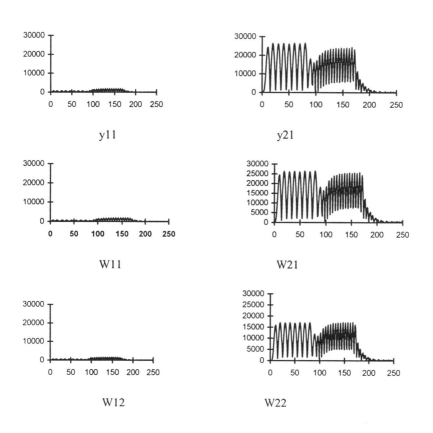

Fig. 7.16: Transposed form IIR LP filter,
unscaled sine input (± 50), f/fs = 0.05, 0.1, 0.2; graph shows first section and
second section node values

However the maximum input before overflow is very small. The reason for this can be seen if the magnitude of the internal variables at the six nodes is plotted with a sine input. Fig. 7.15 shows the sine input used with three frequencies of f/fs = 0.05, 0.1 & 0.2 and the corresponding output agrees with Fig. 7.9.

The filter clearly works as a low pass filter but the maximum output is achieved for a peak sine input of only 50. The magnitude of the six nodes in the program was therefore calculated (by the TMS320 program) and are plotted in Fig. 7.16.

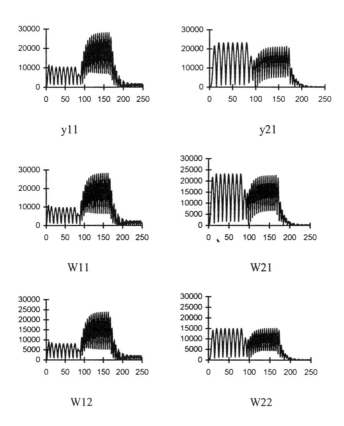

Fig. 7.17: Transposed form IIR LP filter,
scaled / 16, sine input (± 700), f/fs = 0.05, 0.1, 0.2;
graph input shows maximum first section and second section node values

Remembering that, with 16 bit arithmetic, overflow occurs at +32767 and -32768, it can be seen that the first section calculations could handle a much greater dynamic range than they are allowed to by the overflow constraints imposed by the second section computations.

A right shift was carried out on y_{11} when it was used by the second section and the magnitudes again plotted. Naturally, a larger sine input could then be tolerated before overflow. The optimal shift value was found experimentally and found to be a 4-bit shift (/16). The magnitudes at the nodes in the scaled version are shown in Fig. 7.17.

The peak sine input before overflow has now been increased to ±700 from the unscaled version of ±50. This is quite a reasonable method of scaling such a filter, especially as this scaling is not readily solved by theoretical analysis. The input and output of the scaled filter are shown in Fig. 7.18.

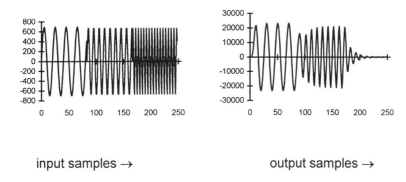

input samples →   output samples →

Fig. 7.18: Transposed form IIR LP filter, y11 scaled /16, sine input (±700), f/fs = 0.05, 0.1, 0.2

This practical approach also demonstrates that it is necessary to look at the overflow at different frequencies, especially in and near the pass band. This particular filter appears to be at its most critical at the turnover frequency. Obviously the same techniques can be used to look at the overflow with inputs other than sine waveforms.

7.8.3 High pass TMS320C25 IIR filter performance

The same program was used to implement the high pass IIR filter with a turnover frequency of f/fs = 0.4, Section 7.7, Eq. (7.33).

The optimal scaling in this case was with y1 scaled by 5 shifts (/32). The response to a sine input in the stop band (f/fs = 0.2) and pass band (f/fs = 0.45) are shown in Fig. 7.19. The dynamic range achieved is of the order of ± 1500 (peak sine). Because the graphs show the sampled input values and the raw output, before filtering, they look rather strange at the very high frequency of f/fs = 0.45, because of the lack of a reconstruction filter to smooth the waveform. However, the filtering action and dynamic range are clearly demonstrated. The pulse response and the maximum values of the node variables are given in Appendix 7.3.

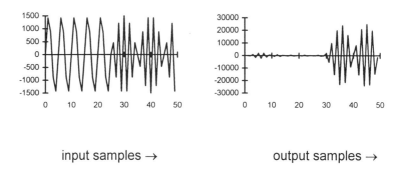

input samples →   output samples →

Fig. 7.19: Transposed form IIR HP filter, y11 scaled /32, sine input (± 700), f/fs = 0.2, 0.45

APPENDIX 7.1: COEFFICIENT QUANTIZATION

The positions of poles and zeros for a filter design with the transfer function H(z), Eqs. (1.25) and (1.27), are important from the point of view of the filter stability (poles) and the shape of the frequency response (poles and zeros).

It can be shown, [27], that the changes in filter coefficients Δa_m cause the changes in pole postions Δp_i

$$\Delta p_i = - \sum_{m=1}^{N} \frac{p_i^{N-m}}{\prod_{\substack{j=1 \\ j \neq i}}^{N} (p_i - p_j)} \Delta a_m \quad ,$$

where $p_i - p_j$ are vectors in the z-plane from the poles $\{p_j\}$ to the pole p_i. The poles are closely clustered, for high order filters, so that the terms $(p_i - p_j)$ can be very small and the reciprocal values are large. Hence small changes in the coefficients a_m, m = 1,...N, can cause large changes in the poles of the filter and consequently change the transfer function H(z).

The error Δp_i can be minimised by maximising the lengths $(p_i - p_j)$ which is achieved by using single-pole or double-pole filter sections.

An analogous result can be obtained for the sensitivity of zeros to errors in parameters $\{b_m\}$. Therefore to reduce the sensitivity to coefficients variation a high order filter should be implemented as a combination of first and second order sections in cascade or parallel structure.

As mentioned above, the sensitivity analysis performed on the poles also applies directly to zeros of the IIR filters, hence also to zeros of the FIR filters. Therefore, we should generally realise FIR filters with a large number of zeros as a cascade of second-order and first-order sections in order to minimise the sensitivity to coefficient quantization, [27].

In general, rounding the coefficients of the difference equation to (b + 1) bits will result in errors in the positions of the poles and zeros of the transfer function. The effect of coefficient rounding can be determined by comparing the frequency response of the ideal filter implemented with 32-bit coefficients   with that of the actual   filter implemented with quantized coefficients of (b + 1) bits.

APPENDIX 7.2: IIR LP FILTER LISTING FOR THE TMS320C25

(transposed structure, 21 coefficients with scaling)

```
                    ;low pass filter transposed structure
                    ;>-1<+1 arithmetic used throughout
                    ;final fast version using shifting for b11, b12, b21, b22
                    ;original code for unscaled, direct multiply version
                    ;in comments marked ***
                    ;35 machine cycles per sample
                    ;
scale               equ 4     ;y11 scaling factor >>4 bits
;
                    origin 0
                    include 'h32025.inc'
;
input               equ 0    ;input port address
output              equ 0    ;output port address
                    ;define data memory for intermediate results
                    ramaddr  300h   ;on chip RAM base
x0                  ramspace 1
W11                 ramspace 1
W12                 ramspace 1
y11                 ramspace 1
W21                 ramspace 1
W22                 ramspace 1
y21                 ramspace 1
                    ;copy coefficients here for use

a11                 ramspace 1
a12                 ramspace 1
a21                 ramspace 1
a22                 ramspace 1
b11                 ramspace 1
b12                 ramspace 1
b21                 ramspace 1
b22                 ramspace 1
                    ;
                    ;compute algorithm thus
                    ;y11 = x0 + W11
                    ;W11 = 2 * (b11/2 * x0 + -a11/2 * y11 + W12/2)
                    ;W12 = 2 * (b12/2 * x0 + -a12/2 * y11)
                    ;y21 = y11 + W21
                    ;W21 = 2 * (b21/2 * y11 + -a21/2 * y21 + W22/2)
                    ;W22 = 2 * (b22/2 * y11 + -a22/2 * y21)
                    ;
                    ;copy coefficients to data memory
                    larp ar1
```

```
                lrlk ar1,a11
                rptk coeflen - 1
                blkp coefstart,*+
                ldpk h.x0  ;point to 300h to 37fh
                spm pshift1    ;work in acc b16-31 after multiply
    .
    ;
    loop        ;start first section
                in l.x0,input
                zalh l.x0
                addh l.W11        ; acch = x0+W11
                sach l.y11,noshift ;  -> y11
                                 ; lt l.x0 ***
                zalh l.x0        ; mpy l.b11 ***
                           ; pac ; acch = x0*b11/2 ***
                lt l.y11
                mpy l.a11
                apac ; + y11*a11/2
                add l.W12,shift15 ; + W12/2
                sach l.W11,shift1 ; * 2 -> W11
                                 ; lt l.x0 ***
                lac l.x0,shift15  ; mpy l.b12 ***
                               ; pac ; acch=x0*b12/2 ***
                lt l.y11
                mpy l.a12
                apac              ; + y11*a12/2
                sach l.W12,shift1 ; *2 -> W12
    ;           start second section
                           ; zalh l.y11  ***
                lac l.y11,(16-scale)
                addh l.W21        ; acch = y11+W21
                sach l.y21,noshift  ; -> y21
                ;now do acch = (y11*b21/2) >> 4 bits scaling
                              ; lt l.y11 ***
                lac l.y11,(16-scale) ; mpy l.b21 ***
                             ; pac ; acch = y11*b21/2 ***
                lt l.y21
                mpy l.a21
                apac  ; + y21*a21/2
                add l.W22,shift15  ; + W22/2
                sach l.W21,shift1  ; *2 -> W21
                ;now do acch = (y11 * b22/2) >> 4 bits scaling
                             ; lt l.y11 ***
                lac l.y11,(15-scale); mpy l.b22
                             ; pac ; acch=y11*b22/2 ***
                lt l.y21
                mpy l.a22
                apac              ; + y11*a12/2
                sach l.W22,shift1 ; *2 -> W22
                out l.y21,output
```

```
                b loop
    ;
                ;define filter constants in range  -1 to +1
coefstart
                data  24583 ; 0.7502 = a11/2
                data -13889 ; -0.4239 = a12/2
                data  25471 ; 0.7773 = a21/2
                data  -10640  ; -0.3247 = a22/2
                data  32767 ; 1.0 = b11/2
                data  16384 ; 0.5 = b12/2
                data  32767 ; 1.0 = b21/2
                data  16384 ; 0.5 = b22/2
coeflen         equ $ - coefstart
```

APPENDIX 7.3: HP TMS320C25 IIR FILTER RESPONSE

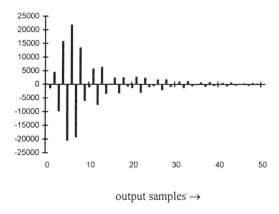

output samples →

A7.1: Impulse response (6000), transposed form IIR HP filter, y11 scaled /32

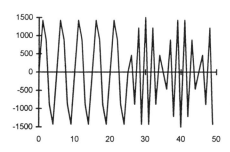

input samples →

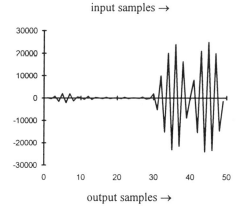

output samples →

Fig A7.2: Transposed form IIR HP filter, y11 scaled /32 sine input (± 700), f/fs = 0.2, 0.45

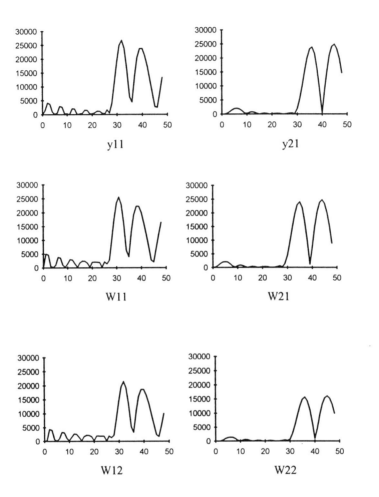

Fig. A7.3: Transpose form IIR HP filter, scaled /32,
sine input ± 700), f/fs = 0.2, 0.45;
graphs show maximum first section and second section
node values during the execution of Fig. A7.2

8

Other topics in digital signal processing

8.0   INTRODUCTION

The topics presented in this chapter are more specialised than those in previous chapters. However, they are widely used in many applications such as communications, control, geophysics etc. They usually appear in more advanced textbooks on digital signal processing, but they are covered here in an introductory and tutorial manner.

In this first section, adaptive FIR type filters are introduced with a simplified derivation of the LMS algorithm. The following Section 8.2 deals with concepts of inverse filtering, deconvolution and system identification. A more specialised form of deconvolution, called homomorphic deconvolution, is introduced in Section 8.3. In Section 8.4, the forward linear prediction concepts are developed using a simplified approach. In the following Section 8.5, a simple control system is discussed with relevant design features. This is followed by a design example showing various design styles. Finally, in Section 8.6, the design examples of an adaptive filter and a digital control are implemented on the TMS320C25/C50.

8.1   ADAPTIVE FILTERS

The main difference between the adaptive signal processing methods and usual signal processing, described in previous chapters is that we are now dealing with time-varying digital systems. Adaptive signal processing is applied in many fields such as radar and sonar systems, speech analysis, seismic work and similar other areas. Also, the work can be extended from the single-input to the multiple-input adaptive arrays.

The basic adaptive system is shown in Fig. 8.1, where the single input and output sequences are $x(n)$, and $y(n)$, but the transfer function describing their

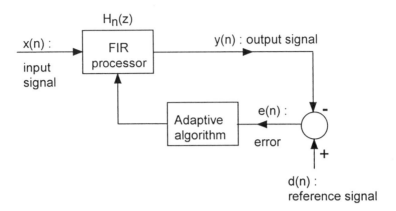

Fig 8.1: Block diagram of an adaptative system

relationship $H_n(z)$ is now time-varying. We consider the case of a time-varying FIR filter described by the following convolution equation

$$y(n) = \sum_{m=0}^{N-1} w_m(n)\, x(n-m) \qquad .$$ (8.1)

By comparison with the time-invariant convolution, Eq. (1.29), $h(m)$ is changed to $w_m(n)$. This is the usual notation for adaptive filters, showing also the time dependence of FIR coefficients by the variable n. It is assumed that the desired (or reference) signal $d(n)$ is available. The objective is to find the best impulse response set of FIR filter coefficients $\{w_m\}$. The 'best' is defined to mean the choice of $\{w_m\}$ that minimises the mean square error (MSE):

$$E\{|e(n)|^2\} = E\{|d(n) - y(n)|^2\} \qquad ,$$ (8.2)

where $e(n)$ is the error signal and $y(n)$ is given by Eq. (8.1).

For the FIR case, the MSE is a quadratic function of the filter coefficients and hence has a single minimum. This is illustrated in Fig. 8.2 for one coefficient, where $w_0(0)$ denotes the initial condition, and w_0^* corresponds to the optimal, i.e. the minimum mean-square error (MMSE) solution. In the general case, the MSE is a multidimensional parabolic function which is more difficult to visualise.

The aim of the adaptive process is to adjust the filter coefficients so that they move from any initial set of values $\{w_n(0)\}$, towards MMSE solution set $\{w_n^*\}$. In nonstationary situations, the MMSE solution varies as signal conditions

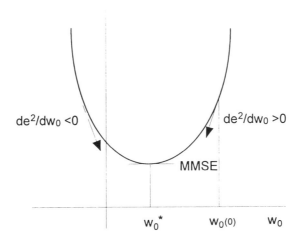

Fig 8.2: Mean square error (MSE) quadratic function with minimum MMSE at w_0^*

change. Therefore, the adaptive process must continually adjust the coefficients in order to track the MMSE solution. However, in adaptive system analysis, we assume stationary, i.e. very slow change in signal statistics. The simplest and perhaps most frequently used adaptive procedure is given by the least-mean square (LMS) algorithm.

This algorithm for updating filter coefficients is based on the method of steepest descent from an initial set of values to the optimum MMSE values [13, 32]. It is described by the vector equation

$$\underline{w}(n+1) = \underline{w}(n) + \mu(-\underline{\nabla}_n) \quad , \qquad (8.3)$$

where

$$\underline{w}(n) = \begin{bmatrix} w_0(n) \\ \vdots \\ w_{N-1}(n) \end{bmatrix}, \text{ and } \quad \underline{\nabla}_n = \begin{bmatrix} \partial E(e_n^2) / \partial w_0(n) \\ \vdots \\ \partial E(e_n^2) / \partial w_{N-1}(n) \end{bmatrix}$$

are the coefficient vector and MSE gradient vector respectively; and μ is a parameter controlling the rate of convergence. Note that the coefficient updating is proportional to the negative gradient $(-\underline{\nabla}_n)$. Hence, as indicated in Fig. 8.2, the MSE will decrease from the initial value on either side of the minimum. Also, once the MMSE is reached, the gradient becomes zero and so the coefficients remain at their optimal values.

The algorithm given by Eq. (8.3) is not suitable for practical implementation because of inexact knowledge of the gradient ∇_n. There are various techniques for estimating ∇_n, but the LMS approach uses the estimate based on the instantaneous squared error

$$\hat{\nabla}_n = \frac{\partial e^2(n)}{\partial \underline{w}(n)} = 2e(n)\frac{\partial[d(n) - y(n)]}{\partial \underline{w}(n)} \quad , \qquad (8.4)$$

where e(n), d(n) and y(n) are as shown in Fig. 8.1. The desired response d(n) is independent of the filter coefficients. The output derivative $\partial y(n)/\partial \underline{w}(n)$ applied to Eq. (8.1) produces $\underline{x}(n)$, so that

$$\hat{\nabla}_n = -2e(n)\underline{x}(n) ,$$

where

$$\underline{x}(n) = \begin{bmatrix} x(n) \\ x(n-1) \\ \vdots \\ x(n-N+1) \end{bmatrix} \qquad (8.5)$$

is a vector of input signal values. Using the above in Eq. (8.3), we obtain

$$\underline{w}(n+1) = \underline{w}(n) + 2\mu e(n)\underline{x}(n) \quad , \qquad (8.6)$$

where μ is the convergence parameter, e(n) is the error signal

$$e(n) = d(n) - y(n)$$

and $\underline{x}(n)$ is the input signal vector as defined in Eq. (8.5). The parameter μ plays an important role in determining the performance of an adaptive system. A large μ could result in an adaptive process that never converges to the MMSE solution. But, if μ is too small, the coefficient vector adaptation is very slow.

The theoretical studies [13, 32] establish that the adaptive constant μ is bounded by

$$0 < \mu < \frac{2}{N}\frac{1}{\sigma^2} \quad , \qquad (8.7)$$

where N = number of filter coefficients, and σ^2 is the input signal power.

The **LMS algorithm** is applied repeatedly as follows: calculate the output y(n), Eq. (8.1), the error e(n) = d(n) - y(n), and then update FIR weights Eq. (8.6), as summarised below:

$$
\begin{aligned}
y(n) &= \underline{w}'(n)\underline{x}(n) &&\text{(output)}\\
e(n) &= d(n) - y(n) &&\text{(error)}\\
\underline{w}(n+1) &= \underline{w}(n) + \mu\, e(n)\underline{x}(n) &&\text{(weights update)}
\end{aligned}
\tag{8.8}
$$

In these equations $\underline{w}(n)$ is the weights column vector, $\underline{w}'(n)$ is its transpose, i.e. the row vector.

A number of interesting applications of the adaptive signal processing can be found in [13, 32].

Example 8.1

To illustrate the operation of the LMS algorithm we consider a communication system, Fig. 8.3, in which the communication channel (C) is followed by the adaptive filter situated in a receiver. The input to the channel is a repeated sequence of unit pulses $\{\delta(n)\}$ in well spaced intervals between which the adaptive filter can reach the steady state.

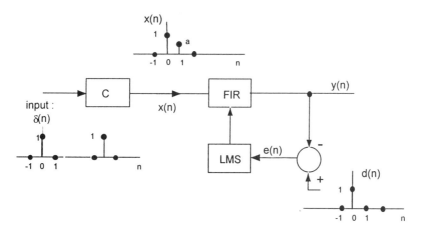

Fig. 8.3. A simplified communication system, with periodic unit-sample input and output x(n).
The receiver side contains a LMS adaptive system.

As shown in Fig. 8.3, the channel pulse response has the main response $x(0) = 1$ and an undesirable additional pulse $x(1) = a$. The adaptive filter is to process the incoming sequence, with the aim of removing $x(1) = a$ in its output, by means of the LMS algorithm given by the set of equations (8.8).

We consider FIR with two weights (N=2), and assume the initial weights $\underline{w}'(0) = [K_0 K_1]$, where K_0, K_1 are arbitrary constants.

The input vector is

$$
\underline{x}(n) = \big[x(n)\,x(n-1)\cdots x(n-N+1)\big]', \text{ for } N = 2 \text{ is } \underline{x}(n) = \big[x(n)\,x(n-1)\big]'.
$$

The desired, or reference signal is $d(n) = \delta(n)$.
The value for the parameter μ, Eq. (8.7), is given by

$$\mu = 1/\sigma^2 \ ,$$

where $\quad \sigma^2 = 1 + a^2 \cong 1$, assuming $a \ll 1$; therefore we use $\mu = 1$.

Solution

The first set $\{x(n)\} = \{1,a\}$ is to be processed by means of the LMS algorithm Eq. (8.8) for times $n = 0, 1,2,3$. The initial FIR weights are $w_0(0) = K_0$, $w_1(0) = K_1$. Also, note that $x(n) = 0$ for $n < 0$.

		(1)	(2)	(3)
Initial { K_0		1	0.25	0.9375
Weights { K_1		1	- 0.375	- 0.375
$y(0) = K_0 \times (0) + K_1 x(-1) = K_0$		1	0.25	0.9375
$e(0) = d(0) - y(0) = 1 - K_0$ $w_0(1) = w_0(0) + e(0) x (0) = K_0 + 1 - K_0 = 1$ $w_1(1) = w_1(0) + e(0) x (-1) = K_1$ $y(1) = x(1) + K_1 x (0) = a + K_1$		1.5	0.125	0.125
$e(1) = d(1) - y(1) = 0 - (a + K_1)$ $w_0(2) = w_0(1) + e(1) x (1) = 1 - a^2 - aK_1$ $w_1(2) = w_1(1) + e(1) x (0) = - a$ $y(2) = (1 - a^2 - aK_1) x (2) - a x(1) = - a^2$		- 0.25	- 0.25	- 0.25
$e(2) = d(2) - y(2) = a^2$ $w_0(3) = w_0(2) + e(2)x(2) = 1 - a^2 - aK_1$ $w_1(3) = w_1(2) + e(2) x(1) = - a + a^3$ $y(3) = 0, \qquad e(3) = 0$		0	0	0
$w_0(4) = w_0(3)$	Final { w_0	0.25	0.9375	0.9375
$w_1(4) = w_1(3)$	Weights { w_1	- 0.375	- 0.375	- 0.375

Table 8.1

Results of successive applications of the LMS algorithm Eq. (8.8) are given on the left hand side of Table 8.1. The steady-state has been reached at $n = 3$. This has been done for the first period of the repetitive input in Fig. 8.3. The procedure continues for the second period of the input signal but now using $w_0(4)$ and $w_1(4)$ weights as the initial FIR weights. One can continue until results do not change any more, Fig. 8.4.

The right hand side of Table 8.1 contains numerical results for the case of $a = 0.5$, and $K_0 = K_1 = 1$. It has three columns (1), (2) and (3) for three periods of the input signal. The final weights of one processing cycle become the initial weights of the next cycle.

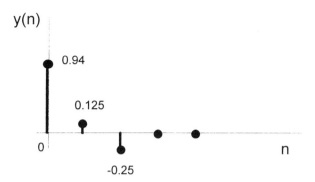

Fig. 8.4: Adaptive filter output y(n) for 2-weights FIR filter, after 3 processing cycles (a = 0.5)

It is interesting to examine the same procedure for the FIR filter with three coefficients. In this case we have $N = 3$, initial weights settings $\underline{w}(0) = \begin{bmatrix} K_0 & K_1 & K_2 \end{bmatrix}'$ and the input samples set $\underline{x}(n) = \begin{bmatrix} 1 & a & 0 \end{bmatrix}'$.

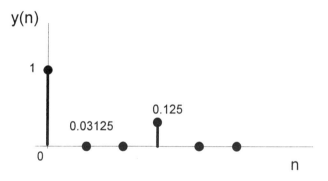

Fig 8.5: As in Fig. 8.4 but now with 3-weights filter

After processing the fourth input set, we obtain the output sequence as shown in Fig. 8.5 and the FIR weights:

$$w_0(4) = 1 - a^6 \simeq 1$$
$$w_1(4) = -a$$
$$w_2(4) = a^2 - a^4 \simeq a^2 \quad for \quad a \ll 1$$

Matlab program for Example 8.1

```
a = 0.5;
K0 = 1; K1 = 1;
x = [ 0   1   a   0   0   0   0 ];
d = [ 1   0   0   0   0   0   0 ];
mu = 1; % mu = μ (in the text)
w = [ K0   K1 ];  % initial two weights FIR
for      n = 1 : 6
         p = [ x (n+1)   x(n) ];
         y = w * p';
         e = d (n) - y;
         q = w + mu * e * p;
         w = q;
         y          % output sequence
end

         w          % final weights
```

(**Note**: Matlab vectors are row vectors, but in the LMS algorithm Eq. (8.8) we have column vectors, hence differences in notation.) Applying the above program in succession we obtain numerical results in agreement with Table 8.1, columns (1), (2), (3).

Comments:

The adaptive filter used in this example has been operating on a given known communication channel. The filter weights have been set up for the sequence of well spaced input pulses. Such an operation is often called the training of an adaptive filter. If the channel response changes slowly with time, the adaptive filter will follow these changes and adjust its coefficients according to the LMS algorithm; Eq. (8.8).

Example 8.2

Consider the adaptive system in Fig. 8.6 where

$$x_1(n) = A \sin \omega_1 n = 4 \sin (2\pi \ 5/32) \ n \quad ,$$
$$x_2(n) = B \sin \omega_2 n = 3 \sin (2\pi \ 9/32) \ n \quad ,$$
and v_1 is Gaussian white noise.

The desired (or reference) signal,
$$d(n) = 2 \sin (\omega_2 n + v_2) \quad ,$$

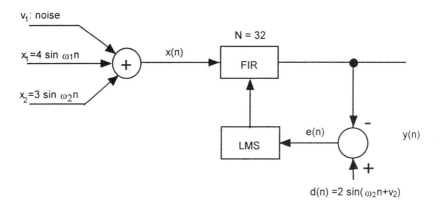

Fig 8.6: Block diagram with for example with $\omega_1 = 2\pi(5/32)$, $\omega_2 = 2\pi(9/32)$,
v_1 is Gaussian noise and v_2 is uniform noise
(both are specified in the Matlab program, see Example 8.2)

is a sinusoid of the same frequency as the signal $x_2(n)$ but with uniform random phase v_2. The FIR filter is taken to have $N = 32$ coefficients (or weights).

The objective is to use the adaptive LMS algorithm Eq. (8.8) in order to extract the signal $x_2(n)$ at frequency ω_2, remove the signal x_1 and suppress the input noise over the whole band. The adaptive constant μ is bounded by

$$0 < \mu < \frac{2}{N} \min\left(\frac{2}{A^2}, \frac{2}{B^2}\right) \quad , \qquad (8.9)$$

which in our case for $A = 4$ and $N = 32$ gives $\mu < 0.0078$.

The operation of this adaptive system has been examined in the following matlab program with the adaptive constant μ denoted as mu = 0.002. The adaptive LMS algorithm is within the *for* loop. We note again that Matlab vectors are row vectors while in Eq. (8.8) we have column vectors.

The program enables the plotting of various waveforms for checking the operation and results with plot (x), plot (v_2) plot (r_1), plot (r_2) and plot (v * 0.5/pi, abs (h)). The last one, given here in Fig. 8.7, shows that the system operates as a bandpass filter at the frequency ω_2. The theoretical details for this example can be found in [32, section 4.4].

Matlab program for Example 8.2

```
n = 0 : 255 ; p1 = 1.5 ; p2 = 0.5 ;
x1 = 4 * sin (( 2 * pi * 5 / 32 ) * n ) ;
x2 = 3 * sin (( 2 * pi * 9 / 32 ) * n ) ;
v1 = p1 * randn ( 1,256 ) ; % Gaussian noise
x = x1 + x2 + v1;                              % input data + noise
plot ( x ) ; grid ; pause
v2 = p2 * ( rand ( 1,256 ) - 0.5 ) ;       % uniform noise
plot ( v2 ) ; grid ; pause
d = 2 * sin (( 2 * pi * 9 / 32 ) * n + v2 ( n + 1)) ;
w = zeros ( 1 , 32 ) ; mu = 0.002 ; % w = initial values
r1 = zeros ( 1, 32 ) ;
for n = 1 : 32               % LMS loop
        p = [ x (( n + 31 ) : - 1 : n ) ] ;
        y = w * p' ;
        e = d ( n ) - y ;
        q = w + mu * e * p ;
        w = q ;
        r1 ( n ) = y ;
end
plot ( r1 ) ; grid ; pause
i = 1 : 32
r2 = sin (( 2 * pi * 9 / 32 ) * i ) ;
plot ( i, r2 ) ; grid ; pause
[ h , v ] = freq z ( w, 512 ) ;
plot (v * 0.5 / pi, abs ( h ) ; grid
```

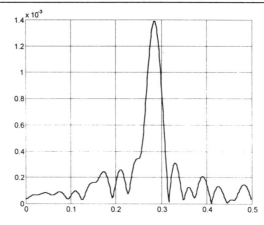

Fig 8.7: Magnitude of the frequency response for the final FIR filter in the adaptive system in Fig. 8.6. The horizontal scale is frequency, given by $\omega/2\pi$; The peak response is at $\omega_2 = 2\pi(9/32)$, which is at $f = 9/32 = 0.28$.

8.2   DECONVOLUTION AND SYSTEM IDENTIFICATION

As shown in Eqs. (1.29) and (4.9) the output of a linear time-invariant system is the convolution of the input signal x(n) with the unit-sample response, h(n), of the system. Often, in practice, an input signal x(n) is a sum of a desired signal, and an undesired interference signal. The system is then designed to suppress the interference but preserve the desired signal. This is the filter design problem treated in Chapters 6 and 7.

Another class of practical problems arises, for example, in high data rate digital communications over telephone channels. The channel distorts the signal by causing binary data symbols spread into adjacent symbol slots, which is known as the intersymbol interference. In such cases the problem is solved by using a corrective system at the receiver which removes the distortion caused by the channel and produces the correct symbol at the output, Fig. 8.8.

One form of the corrective network is the adaptive filter, treated in the previous section, which is used for slowly time varying channels. If the channel characteristics (unit-sample response or frequency response) are known and are not varying, the corrective network is known as the equaliser.

In the linear systems theory, the corrective system is called an inverse system. Since the distortive system, c(n), in Fig. 8.8 yields an output u(n) that is the convolution of the input x(n) with c(n), the inverse system operation that takes

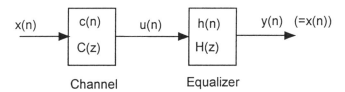

Fig 8.8: Channel (producing convolution) and equaliser (producing deconvolution).

u(n) and produces x(n) is called deconvolution. For the system in Fig. 8.8 we can write the convolution equation.

$$u(n) = x(n) * c(n) \quad .$$

The output of the corrective network is given by the second convolution equation

$$y(n) = u(n) * h(n)$$

$$= x(n) * c(n) * h(n) \quad . \tag{8.10}$$

The deconvolution, $y(n) = x(n)$, is achieved if

$$c(n) * h(n) = \delta(n) \quad , \tag{8.11}$$

where $\delta(n)$ is the unit-sample as defined in Appendix 1.3. The equation can be used to solve for $h(n)$ for a given $c(n)$. However, the solution of Eq. (8.11) in the time domain is difficult. It is simpler to apply the z-transformation to Eq. (8.11) which produces

$$C(z) . H(z) = 1 \quad . \tag{8.12}$$

Therefore, the corrective network transfer function is given by

$$H(z) = C^{-1}(z) \quad , \tag{8.13}$$

which is the inverse of the channel transfer function. If $C(z) = B(z)/A(z)$, i.e. a rational system function, then the inverse is $H(z) = A(z)/B(z)$. Hence, the zeros (poles) of $C(z)$ become poles (zeros) of $H(z)$. A stable pole-zero system that is minimum - phase has a stable inverse which is also minimum phase (i.e. all zeros inside the unit circle).

Example 8.3

The channel unit-sample response of a system is given by

$$c(n) = 0.5^n, n \geq 0 \quad .$$

Determine the inverse for this case.

Solution

The z-transform for given $c(n)$ is

$$C(z) = \frac{1}{1 - 0.5z^{-1}} \quad .$$

This is an all-pole system, causal and stable. Therefore, its inverse gives the corrective network

$$H(z) = C^{-1}(z) = 1 - 0.5 z^{-1} \cdot$$

which is an FIR with the unit-sample (impulse) response

$$h(n) = \delta(n) - 0.5\ \delta(n\text{-}1)\ \ .$$

Example 8.4

Determine the inverse of a communication system with the channel impulse response

$$c(n) = \delta\ (n) + a\ \delta(n\text{-}1)\ \ \ .$$

Solution

This is an FIR system and its transfer function is

$$C(z) = 1 + az^{-1}\ \ .$$

The inverse (corrective) system is

$$H(z) = C^{-1}(z) = \frac{1}{1 + az^{-1}}\ \ ,$$

or after long division

$$H(z) = 1 - a\ z^{-1} + a^2\ z^{-2} - a^3\ z^{-3} + \ldots$$

The impulse response is

$$h(n) = \delta(n) - a\ \delta(n\text{-}1) + a^2\ \delta(n\text{-}2) - a^3\ \delta(n\text{-}3) + \ldots$$

(**Note** : This is the case considered in Example 8.1, where $\underline{w}(4)$ results agree with the appropriate terms in the above h(n). The corrective network there is an adaptive filter whose coefficients are denoted by w, while here they are denoted by h).

In some practical applications the impulse response c(n) may not have a z-transform in closed form. Then assuming that c(n) and h(n) are causal, Eq. (8.11) can be solved directly [27, Chapter 5] as follows:

$$h(0) = 1/c\ (0) \tag{8.14}$$

and
$$h(n) = \sum_{k=1}^{n} \frac{c(n)h(n-k)}{c(0)}, \quad n \geq 1 \tag{8.15}$$

This recursive relation can easily be programmed on a digital computer.

We can determine the impulse response of an unknown system by using a known input sequence, $x(n)$, and observing the output sequence $y(n)$. This is the problem of system identification, with the solution referred to as deconvolution. With reference to Fig. 8.8, we have

$$u(n) = c(n) * x(n) \tag{8.16}$$

and hence

$$C(z) = \frac{U(z)}{X(z)}. \tag{8.17}$$

Example 8.5

For the input sequence

$$x(0) = 1,\ x(1) = -0.9,\ x(2) = 0.\ 2,\ x(n > 2) = 0 \quad,$$

a casual system produces the output

$$u(0) = 1,\ u(1) = 0.7,\ \text{and}\ u(n > 1) = 0 \quad.$$

Determine its impulse response.

Solution

Forming z-transforms and using Eq. (8.17), we obtain

$$C(z) = \frac{1 + 0.7\,z^{-1}}{1 - 0.9\,z^{-1} + 0.2\,z^{-2}} = \frac{1 + 0.7\,z^{-1}}{\left(1 - 0.5\,z^{-1}\right)\left(1 - 0.4\,z^{-1}\right)}$$

and after a partial-fraction expansion we have

$$C(z) = \frac{3}{1 - 0.4\,z^{-1}} - \frac{2}{1 - 0.5\,z^{-1}}.$$

Using Table 1.1 in Section 1.3, we obtain the time domain solution as

$$c(n) = 3 \ (0.4^n) - 2 \ (0.5^n) \quad,$$

which is the impulse response of the unknown network.

In practice the response $u(n)$ is very likely to be infinite in duration, so the above approach would not be useful. The alternative is to deal with Eq. (8.16) directly in the time-domain. For a causal system, the recursive solution is

$$c(n) = \frac{u(n) - \displaystyle\sum_{k=0}^{n-1} c(k)x(n-k)}{x(0)} \quad, \ n \geq 1 \quad\quad (8.18)$$

$$c(0) = u(0)/ \ x(0) \quad.$$

However, when $c(n)$ is of infinite duration, it will be necessary to truncate this solution at some stage.

Other methods, often used in practice for system identification, are cross correlation techniques and also least-squares. In the least-squares method a FIR or IIR model is postulated whose parameters are determined by minimising the square of the error, between the actual system response and the response of the model. The least-squares method is also used to determine FIR (or IIR) inverse filters to a known system. [27, Chapter 8].

8.3   HOMOMORPHIC DECONVOLUTION

As before, we denote $y(n)$ as the output sequence of a linear time-invariant system and $x(n)$ as the input sequence. Then, we can write in the z-domain

$$Y(z) = X(z) \ H(z) \quad,$$

where $H(z)$ is the system transfer function. The logarithm of $Y(z)$ is

$$
\begin{aligned}
\hat{Y}(z) \ &= \ \log Y(z) \\
&= \ \log X(z) + \log H(z) \\
&= \ \hat{X}(z) + \hat{H}(z)
\end{aligned}
\quad\quad (8.19)
$$

We now formally define the so-called complex cepstrum of the sequence $y(n)$ as the sequence $\hat{y}(n)$, which is the inverse transforms of $\hat{Y}(z)$.

Within the region of convergence, $\hat{Y}(z)$ may be presented by

$$\hat{Y}(n) = \sum_{n=-\infty}^{\infty} \hat{y}(n) z^{-n} \quad, \tag{8.20}$$

where

$$\hat{y}(n) = \frac{1}{j2\pi} \oint \hat{Y}(z) z^{n-1} \, dz \quad. \tag{8.21}$$

Eqs. (8.20) and (8.21) are respectively the z-transform of the cepstrum sequence $\hat{y}(n)$, and the inverse z-transform of $\hat{Y}(z)$.

Expressing Eqs. (8.20) and (8.21) on the unit circle ($z = e^{j\omega}$), we obtain the Fourier transform pair

$$\hat{Y}(\omega) = \log Y(\omega) = \sum_{n=-\infty}^{\infty} \hat{y}(n) e^{-j\omega n} \tag{8.22}$$

$$\hat{y}(n) = \frac{1}{2\pi} \int_{-\pi}^{\pi} \hat{Y}(\omega) e^{j\omega n} \, d\omega \quad. \tag{8.23}$$

Applying Eq. (8.21) i.e. taking the inverse z-transform of Eq. (8.19) we obtain

$$\hat{y}(n) = \hat{x}(n) + \hat{h}(n) \quad. \tag{8.24}$$

This result shows that the convolution of two sequences in the time domain has been converted, by the above procedure, into a summation of the cepstrum sequences.

In some applications, such as seismic signal processing and speech signal processing, the characteristics of the cepstral sequences $\hat{x}(n)$ and $\hat{h}(n)$ are sufficiently different so that they can be separated in the cepstral domain. Then the complex cepstrum can be "filtered" to remove one or the other of the convolved sequences. Since the phase information is retained, see Example 8.6, the complex cepstrum is invertible. Therefore, if $\hat{h}(n)$ is rejected from $\hat{x}(n)$ by "filtering", then $\hat{y} = \hat{x}$, and we may then z-transform, exponentiate, and inverse z-transform to obtain the sequence x(n), i.e. x(n) and h(n) have been deconvolved. The process of separating two convolved signals is called deconvolution. The use of the complex cepstrum to perform the separation is called homomorphic deconvolution.

A complete homomorphic deconvolution system is shown in Fig. 8.9. In place of the z-transform and inverse z-transform, DFT is used. More information can be found in [8 & 27].

Example 8.6

Expressing $X(\omega)$ in terms of its magnitude and phase

$$X(\omega) = |X(\omega)| e^{j\theta(\omega)} \quad ,$$

then

$$\hat{X}(\omega) = \log X(\omega) = \log |X(\omega)| + j\theta(\omega) \quad .$$

The complex cepstrum sequence, as in Eq. (8.23) is given by

$$\hat{x}(n) = \frac{1}{2\pi} \int_{-\pi}^{\pi} \left[\log |X(\omega)| + j\theta(\omega) \right] e^{j\omega n} \, d\omega \quad .$$

Therefore the real and imaginary part of $\hat{x}(n)$ are the inverse Fourier transforms of

$$\log |X(\omega)| \text{ and } \theta(\omega).$$

In some applications, for example in speech signal processing, only the real part of the complex cepstrum is computed, while the phase $X(\omega)$ is ignored. This means the sequence $x(n)$ cannot be recovered from real part $\hat{x}(n)$, i.e. the transformation from $x(n)$ to real of $\hat{x}(n)$ is not invertible.

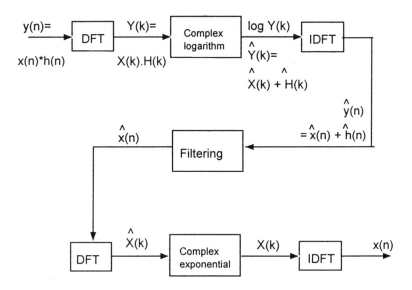

Fig. 8.9: A complete homomorphic deconvolution system also indicating results of processing stages and the removal of h(n)

8.4   LINEAR PREDICTION

In this section, we consider the problem of linearly predicting the value of a time series. This type of processing is applied in a variety of areas such as neurophysics, geophysics, speech communication [17]. The fundamental idea is based on a linear system model driven by some input to produce the output with time specified properties.

A general model for this purpose is described by the difference equation:

$$x(n) = G\sum_{m=0}^{q} b_m\, u(n-m) - \sum_{m=1}^{p} a_m\, x(n-m) \quad . \qquad (8.25)$$

This is the equation we have met in Chapter 1, Eq. (1.24), but with the input now denoted by Gu(n), where G is a gain parameter, the output is denoted by x(n), and the upper limits are p, q. It follows then that the transfer function H(z) = X(z)/U(z) is:

$$H(z) = G\frac{1 + \displaystyle\sum_{m=1}^{q} b_m\, z^{-m}}{1 + \displaystyle\sum_{m=1}^{p} a_m\, z^{-m}} \quad , \qquad (8.26)$$

where we have taken $b_0 = 1$ which is the usual convention. This transfer function is often called the general pole-zero model. Two special cases are of interest:

(i)   all-zero model, $a_m = 0$, $1 \le m \le p$
(ii)   all-pole model, $b_m = 0$, $1 \le m \le q$

In statistical literature, the all-zero model is known as the moving average (MA) model, and the all-pole model is known as the auto regressive (AR) model. In engineering literature, the pole-zero terminology is preferred. We consider, in this section, only the all-pole model.

8.4.1   Forward linear prediction

The difference equation for the all-pole model is:

$$x(n) = -\sum_{m=1}^{p} a_m\, x(n-m) + Gu(n) \quad , \qquad (8.27)$$

where the output x(n) is given as a linear combination of past output values and some input u(n) scaled by a gain factor G, Fig. 8.10.

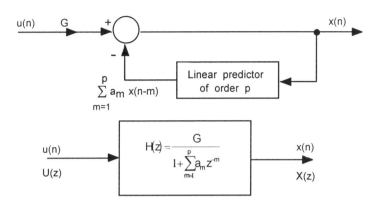

Fig 8.10: Forward linear predictor

The transfer function H(z), Eq. (8.26), now reduces to an all-pole transfer function:

$$H(z) = \frac{G}{1 + \sum_{m=1}^{p} a_m z^{-m}} \tag{8.28}$$

We assume first that in Eq. (8.27) the input u(n) is unknown, which is the case in many applications. Then we consider the one-step forward linear prediction of a signal sample x(n) by a weighted sum of previous samples as:

$$\hat{x}(n) = -\sum_{m=1}^{p} a_m x(n-m) \tag{8.29}$$

The weights $\{-a_m\}$ are called the prediction coefficients of the one-step forward linear predictor of order p. The difference between the value x(n) and its predicted value $\hat{x}(n)$ is called the forward prediction error, denoted as $f_p(n)$ and expressed as:

$$f_p(n) = x(n) - \hat{x}(n)$$

or

$$f_p(n) = x(n) + \sum_{m=1}^{p} a_m x(n-m) \tag{8.30}$$

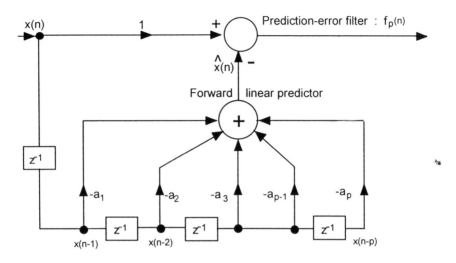

Fig 8.11: One-step prediction and prediction-error formulation

Both of these are shown in Fig. 8.11. The first interpretation of this process is in terms of the forward linear predictor producing $\hat{x}(n)$ and $f_p(n)$. The second interpretation follows from Eq. (8.30), called a prediction-error filter with input x(n) and output $f_p(n)$, which is a direct form FIR filter.

8.4.2   Determination of filter parameters

To simplify finding the solution for filter parameters we assume that x(n) is a deterministic signal, but very similar procedure and results are obtained for a random signal [17].

Denote the total squared error by E, where

$$E = \sum_n f_p^2(n) = \sum_n \left[x(n) + \sum_{m=1}^{p} a_m\, x(n-m) \right]^2 , \qquad (8.31)$$

which we minimise by setting:

$$\frac{\partial E}{\partial a_i} = 0, \qquad 1 \le i \le p.$$

This results in the set of equations:

$$\sum_{m=1}^{p} a_m \sum_n x(n-m)\, x(n-i) = -\sum_n x(n)\, x(n-i) \qquad (8.32)$$

$$i = 1, 2, ..., p ,$$

known as the normal equations. For a signal $x(n)$, Eq. (8.32) forms a set of p equations in p unknowns whose solution gives the predictor coefficients $\{a_m\}$ which minimise E in Eq. (8.31).

The minimum total squared error, denoted by E_p, is obtained by expanding Eq. (8.31) and substituting the solution Eq. (8.32), which results in:

$$E_p = \sum_n x^2(n) + \sum_{m=1}^{p} a_m \sum_n x(n) \times (n-m) \quad . \tag{8.33}$$

If the error in Eq. (8.31) is minimised over the infinite duration, $-\infty < n < \infty$, Eqs. (8.32) and (8.33) reduce to:

$$\sum_{m=1}^{p} a_m r(i-m) = -r(i) \qquad 1 \le i \le p \tag{8.34a}$$

$$E_p = r(0) + \sum_{m=1}^{p} a_m r(m) \quad , \tag{8.34b}$$

where $r(i-m)$, $r(i)$ are the autocorrelation functions. More detail can be found in [6, 17 & 27]. To complete this analysis we rewrite Eq. (8.30) as

$$x(n) = -\sum_{m=1}^{p} a_m x(n-m) + f_p(n) \quad . \tag{8.35}$$

Now comparing Eqs. (8.27) and (8.35) one can see that the only input signal $u(n)$ resulting in $x(n)$ as output is given by $G\,u(n) = f_p(n)$. This means that the input signal is proportional to the error signal. For any other input $u(n)$, the output from $H(z)$ in Fig. 8.10 will be different from $x(n)$. But, if we insist that whatever the input $u(n)$, the energy in the output signal must equal that of the original signal $x(n)$, then we can at least specify that the total energy of the input signal $G\,u(n)$ must equal the total energy in the error signal given by E_p in Eq. (8.33).

Two types of inputs are of special interest: the deterministic impulse and stationary white noise. It can be shown [17] that the autocorrelation coefficients of an all-pole filter are the same whether the input is a signal impulse or white noise. This is very useful in modelling speech process [6] so that unit impulse as well as white noise can be used for speech systems. There are also other interesting topics in linear prediction such as backward linear prediction, use of lattice filter structure for both the forward and backward prediction, and computationally efficient Levinson-Durbin algorithm for solving the normal equation [6, 27].

Example 8.7

A second order all-pole system, driven by a unit pulse, generates the following signal set.

$$\{x(n)\} = \{1, 1.25, 1.19, 0.82, 0.65, 0.50, 0.38, 0.29, 0.22, 0.17\}$$
$$\uparrow$$

(i) Determine parameters a_1, a_2, of the system generating this signal.
(ii) Check that the solution enables one step prediction of the members of the above given signal set.

Solution

(i) To find the solution we use the normal equations (8.34a), with p=2

$$\sum_{m=1}^{2} a_m r(i-m) = -r(i) \qquad , i = 1,2 \quad ,$$

where r $(\cdot)$ are the autocorrelation sequences given by

$$r(l) = \sum_{n=-\infty}^{\infty} x(n) \, x(n-l), \quad l = 0, \pm 1, \pm 2,... \tag{8.36}$$

Expanding the normal equations, for this case we have

$$\begin{array}{ll} a_1 \, r(0) + a_2 \, r(-1) = - \, r(1) & , \\ a_1 \, r(1) + a_2 \, r(0) = - \, r(2) & . \end{array} \tag{8.37}$$

In order to solve for parameters a_1 and a_2, we need to determine r(0), r(1), r(-1) and r(2). These are obtained from Eq. (8.36) by multiplication and summation of sequences $\{x(n)\}$ with $\{x(n)\}$ for r(0), $\{x(n)\}$ $\{x(n-1)\}$ for r(1), etc. The results rounded to two digits are r(0) = 6.66, r(1) = 6.05, r(-1) = 6.07, r(2) = 5.05, for which solutions of Eq. (8.37) are

$$a_1 = - \, 1.24 \text{ and } a_2 = 0.383 \quad .$$

Note that $r(1) \cong r(-1)$ which is correct since the autocorrelation is a symmetric or even function.
(ii) Using Eq., (8.29) with p=2 we have

$$\begin{array}{ll} \hat{x}(n) & = - \quad a_1 x(n-1) - a_2 x(n-2) \\ & = \quad 1.24 \, x(n-1) - 0.383 \times (n-2) \quad . \end{array}$$

Starting with x(0) = 1, and x (-1) = 0, we have

$$\hat{x}(1) = 1.24, \hat{x}(2) = 1.15, \hat{x}(3) = 0.95, \quad \text{etc}.$$

which agree reasonably well with the values in the given set, considering the rounding of the autocorrelation values to two digits.

8.5   CONTROL SYSTEM THEORY AND DESIGN

In this section we consider a typical analogue control system Fig. 8.12(a). The transfer function of a plant is given by H(s), and D(s) is the transfer function of a compensation network designed so that the control system operates correctly. The digital control is shown in Fig. 8.12(b) where D(s) is replaced by D(z) consisting of ADC, DSP and DAC, behaving closely as D(s) in the analogue system. An equivalent block diagram for Fig. 8.12(b), suitable for analysis and design in discrete-time domain, is shown in Fig. 8.12(c) with sampling at intervals T.

The interface between discrete-time and continuous-time of the system is provided by the zero-order hold (ZOH), following DAC, which approximates a continuous signal by holding the signal sample constant during the interval T. The transfer function of the ZOH is given by

$$\text{ZOH}(s) = \frac{1 - e^{-sT}}{s} . \qquad (8.38)$$

Details about this and other types of holds can be found in refs. [15, 21, 22 & 27].

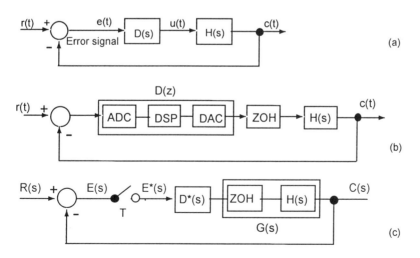

Fig 8.12: (a) Continuous time control system
          (b) Digital control D(z)
          (c) Discrete time equivalent diagram for (b)

The sampled quantities in Fig. 8.12(c) are E*(s), D*(s) and considering the others (H(s), R(s), C(s)) at fictitious sampling times t = kT (k = 0, 1, 2,....) all quantities are in discrete-time domain [22, p.103]. Therefore, we can express all quantities in the z-domain and derive the overall closed-loop transfer function as

$$CL(z) = \frac{C(z)}{R(z)} = \frac{OL(z)}{1 + OL(z)} \quad , \qquad (8.39)$$

where OL(z) = D(z) G(z) is the open-loop transfer function.

The design of a control system is done by adjusting the open-loop frequency response in order to obtain acceptable transient response of the closed-loop system. The main design criteria for this purpose are the open-loop phase margin, gain margin and one of the static error constants [15, 22]. The phase and gain margins are used as a measure of stability by the closeness of the OL(ω) locus to -1 + j0 point. These margins are indicated in the following example.

Example 8.8

Design a control system as in Fig. 8.12 such that the phase margin is about 45°, the gain margin at least 10dB, and the static velocity error constant K_v is 2.3 sec^{-1}. Assume the sampling interval T = 0.2 sec.

The transfer function for the ZOH and the plant H(s) is given by

$$G(s) = \left(\frac{1 - e^{-sT}}{s}\right)\left(\frac{K}{s(s+1)}\right) \quad , \qquad (8.40)$$

where H(s) = K/s(s+1).

The objective is to design the compensation network, D(s), such that the system meets the specification. The design proceeds by first transforming G(s) into G(z) as follows

$$\begin{aligned} G(z) &= Z\left[\frac{1 - e^{-sT}}{s} \frac{K}{s(s+1)}\right] \\ &= K\left(1 - z^{-1}\right) Z\left[\frac{1}{s^2(s+1)}\right] \quad , \end{aligned} \qquad (8.41)$$

where we have used Eq. (1.14), i.e. z = e^{sT}. The expression in the square brackets needs to be expanded into partial fractions:

$$\frac{1}{s^2\,(s+1)} = \frac{A}{s+1} + \frac{Bs+C}{s^2} \quad , \tag{8.42}$$

where $A = -\,B = 1, C = 1$, results in

$$\frac{1}{s^2\,(s+1)} = \frac{1}{s^2} - \frac{1}{s} + \frac{1}{s+1} \quad . \tag{8.43}$$

Applying the transformations from the table in Appendix 8.1, we have

$$Z\left[\frac{1}{s^2\,(s+1)}\right] = \frac{Tz}{(z-1)^2} - \frac{z}{z-1} + \frac{z}{z-e^{-T}} \quad . \tag{8.44}$$

Using Eq. (8.44) in Eq. (8.41) we obtain

$$G(z) = K\,\frac{0.01873\,z + 0.0176}{z^2 - 1.8187\,z + 0.8187} \quad . \tag{8.45}$$

The next step is to transform G(z) to G(w) using the bilinear transformation Eq. (7.3), with the change in notation from s to w and rearrange into

$$z = \frac{1+(T/2)\,w}{1-(T/2)\,w} \quad . \tag{8.46}$$

The notation in control design uses w instead of s in the bilinear transformation. In fact s is used in $z = e^{sT}$ to transform s-plane strips into unit circle, Fig. 1.3. Then, the bilinear transformation w = (2/T) (z-1)/(z+1) is used to transform the z-plane circle into the whole of the l.h.s. of the w-plane. More detail can be found, for example, in [22, Section 4.6].

Substituting Eq. (8.46) into (8.45), and after sorting, we obtain the following result

$$G(w) = K\,\frac{-\,0.000311w^2 - 0.09672w + 0.9988}{w^2 + 0.9969w} \quad . \tag{8.47}$$

The above result can be also obtained using Matlab program 1 for the transformation G(z) to G(w).

Matlab program 1 for Eq. (8.45)/(8.47)

```
% 1st step G(z) : G(-z) = num / den
num = [ 0    - 0.01873   0.0176];
den = [ 1    1.8187    0.8187];
[ numv, denv ] = bilinear (num, den, 0.5);
% v = - (T/2) w = - 0.1w
% num w = [ num v ]. * [0.01 - 0.1  1]
% den w = [ den v ]. * [0.01 -0.1  1]
% den w unity for w^2, multiply num w and den w by 100
num w = [num v]. * [1  -10   100]
den w = [den v]. * [1  -10   100]
```

The compensator transfer function, in w-plane, is assumed to be of the form

$$D\left(w\right) = \frac{1 + \tau\, w}{1 + \alpha\, \tau\, w} \qquad . \qquad (8.48)$$

The static velocity error constant is calculated as follows

$$K_V = \lim_{w \to 0} w\, D\left(w\right) G\left(w\right) = K$$

and since K_v is specified as 2.3 sec^{-1}, we have K = 2.3.

With K = 2.3 we can obtain the frequency response of G(w) by calculating the magnitude and phase by hand or much more easily by using the Matlab *bode* command, which produces plots directly. In the case of our G(w) we would find that the phase and gain margins do not satisfy the specification. Therefore, we need to use the compensation network D(w). It can be shown [21, 22] that for this purpose the compensation network parameters are $\tau = 0.9790$ and $\alpha\tau = 0.3534$, so that

$$D\left(w\right) = \frac{0.9790\, w + 1}{0.3534\, w + 1} \qquad (8.49)$$

Now the open-loop transfer function OL(w) = D(w) G(w), for K = 2.3, is obtained as

$$OL\left(w\right) = \frac{-0.0007\, w^3 - 0.219\, w^2 + 2.027\, w + 2.297}{0.3534\, w^3 + 1.3523\, w^2 + 0.9969\, w} \qquad . \qquad (8.50)$$

Matlab program 2 for Fig. 8.13

```
% OL (w) = num / den : open-loop transfer function
num = [ -0.0007  -0.219  2.027  2.297 ];
den = [ 0.3534  1.3523  0.9969  0 ];
w = logspace (-1, 3, 100);
[ mag, phase, w ] = bode (num, den, w);
magdB = 20 * log 10 (mag);
mg0 = 0* ones (1,100);
subplot (2, 1, 1)
semilog x (w, mag dB, w, mgo); grid
ph 180 = - 180 * ones (1, 100);
subplot x (w, phase, w, ph 180); grid
pause; clf
% Nyguist diagram, polar form
v = 0 : 0.5 : 10;
[ re, im, v ] = nyquist (num, den, v);
z = re + i * im;
r = abs (z) ; theta = angle (z);
polar (theta, r ) ; grid
pause ; clf
```

The Bode plots for Eq. (8.50), obtained from the first part of Matlab program 2, are shown in Fig. 8.13. For 0dB gain crossover point we find the phase margin to be approximately 42°. The gain margin is obtained, for -180° phase crossover point, approximately -13dB.

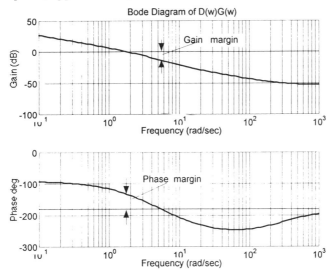

Fig. 8.13: The open-loop response (Bode plot) for example 8.9

The second part of program 2 produces Nyguist polar plot (not shown here). If gives the phase margin of 45°, and the gain margin -14dB, which agree well with Bode plots, but are more precisely determined.

In order to obtain the closed-loop response in discrete-time we transform OL (w), Eq. (8.50), into OL(z) using the bilinear transformation

$$ w = \frac{2}{T} \frac{z-1}{z+1} \quad , \tag{8.51} $$

where fs = 1/T = 1/0.2 = 5. This can be done by hand or by using the Matlab command.

$$ [\text{ numz, denz }] = \text{bilinear (num, den, fs)} \quad , \tag{8.52} $$

where num and den are from Eq. (8.50) and fs = 5.
The result is

$$ OL(z) = \frac{0.1026\,z^2 + 0.0129\,z - 0.0786}{z^3 - 2.3776\,z^2 + 1.8351z - 0.4576} \quad . \tag{8.53} $$

The closed-loop transfer function, Eq. (8.39), with the above OL (z) is given by

$$ CL(z) = \frac{0.1026\,z^2 + 0.0129\,z - 0.0786}{z^3 - 2.275\,z^2 + 1.848\,z - 0.536} \quad . \tag{8.54} $$

The unit-step response for Eq. (8.54) is shown in Fig. 8.14. It is obtained with Matlab program 3, where numc and denc are the numerator and the denominator of Eq. (8.54).

Matlab program 3 for Fig. 8.14

```
% closed-loop unit-step response
numc = [ 0   0.1026   0.0129   -0.0786 ] ;
denc = [ 1   -2.275   1.848   -0.536 ] ;
r = ones (1, 41) ;
axis ( [ 0   40   0   1.6 ] ) ;
k = 0 : 40 ;
c = filter (numc, denc, r ) ;
plot (k, c )
grid
```

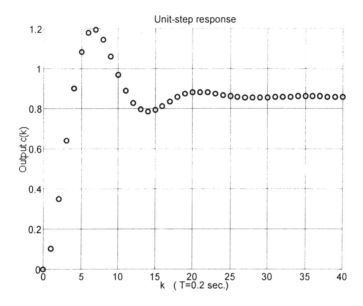

Fig 8.14: The closed-loop response for Example 8.9

8.6   DSP IMPLEMENTATION OF ADAPTIVE FILTER AND DIGITAL CONTROL

This section describes the implementation on the TMS320C25/C50 integer DSP of the adaptive filter in Example 8.2 (Section 8.1) and of the digital control just described.

8.6.1 Adaptive FIR LMS filter (DSP implementation)

The heart of the adaptive filter is the normal DSP convolution operation. A crucial practical consideration is that, as filter coefficients must be updated as the program executes, they must be held in RAM. In the normal MACD convolution instruction coefficients are in program memory; therefore the CNFP instruction must be used (C25) to switch data memory at 200H to 2FFH to program memory at FF00H to FFFFH. The convolution section is written in the usual way and resides in normal off-chip program memory. The first action of the program (listed in full in Appendix 8.2) is to load the convolution + filter coefficients into on-chip data memory. This is switched to on-chip program memory for convolution execution and then back to data memory (CNFD) to allow the efficient updating of the filter coefficients. The whole program was executed on a

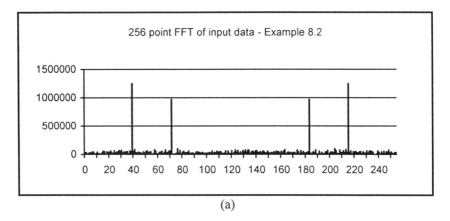

(a)

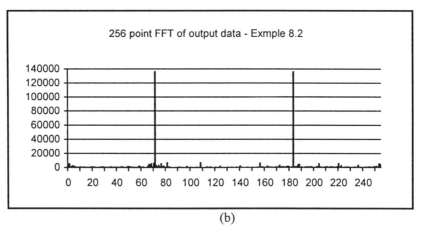

(b)

Fig 8.15: Input (a) and output (b) spectra of DSP implementation of adaptive filter
Example 8.2

TMS320C25 simulator, using input data (x) of two sine waveforms and random
noise: 4 sin (2π5/32)   +   3 sin (2π9/32)   +   noise .

A 256 point FFT of this input data is shown in Fig. 8.15(a) with the sine
waveforms appearing at points 39 and 72. The sin (2π9/32) waveform was used as
the reference input.

The program was executed for 10000 input data samples to allow the filter
coefficient values to stabilise. A FFT of the last 256 output data samples is shown
in Fig. 8.15(b). The 4 sin (2π5/32) waveform has almost disappeared and the
noise is much reduced, just as expected. Note that the integer implementation has
only a limited dynamic range before overflow occurs or truncation errors occur.
The poorer than expected noise reduction is probably due to truncation errors.
The use of a floating point processor would solve this problem. An alternative is

described in [32, Appendix B], where the coefficients are stored and updated in double precision, but the multiply-accumulate operation is carried out only on the most significant half.

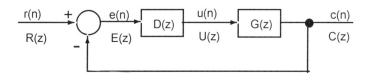

Fig 8.16: Discrete-time form of the control system in Example 8.8

8.6.2 Digital control (DSP implementation)

The discrete time control system considered here is shown in Fig. 8.16 and is fully described by its closed loop transfer function Eq. (8.54). For DSP implementation, we require the time domain description in terms of a difference equation. This is derived from Eq. (8.54) by dividing the numerator and denominator by z^3, so that we have

$$CL(z) = \frac{C(z)}{R(z)} = \frac{0.1026z^{-1} + 0.129z^{-2} - 0.0786z^{-3}}{1 - 2.275z^{-1} + 1.848z^{-2} - 0.536z^{-3}} \qquad . \qquad (8.55)$$

Now we proceed by cross-multiplying, in Eq. (8.55), to obtain

$$c(z) - 2.275(z^{-1}C(z)) + 1.848(z^{-2}C(z)) - 0.536(z^{-3}C(z))$$
$$= 0.1026(z^{-1}R(z)) + 0.129(z^{-2}R(z)) - 0.0786(z^{-3}R(z)) . \qquad (8.56)$$

Applying the inverse z-transform, as discussed in Chapter 1, we have the difference equation for our control system as

$$c(n) = 0.1026\ r(n\text{-}1) + 0.0129\ r(n\text{-}2) - 0.0786\ r(n\text{-}3)$$
$$+ 2.275\ c(n\text{-}1) - 1.848\ c(n\text{-}2) + 0.536\ c(n\text{-}3) \quad . \qquad (8.57)$$

For the unit step response, we have the input step given by
$$r(n) = 1 \text{ for } n \geq 0, \text{ and } r(n) = 0 \text{ for } n<0.$$

Also, the output for n<0 is zero, i.e. $c(-1) = c(-2) = c(-3) = 0$.

We scale Eq. (8.57) for DSP implementation and the algorithm can be described by

Cs ← -0.0345 * r(n-3) + 0.00567 * r(n-2) + 0.0451 * r(n-1)
                              + 0.2356 * c(n-3) + -0.8123 * c(n-2) + c(n-1)
r(n-3) ← r(n-2)
r(n-2) ← r(n-1)
r(n-1) ← r(n)
c(n-3) ← c(n-2)
c(n-2) ← c(n-1)
c(n-1) ← Cs * .5688 * 4   [= C(n)]   .

We have now achieved multiplier constants which are less than 1 and suitable for the normal fractional format in the program. (The *4 is a two bit left shift). A complete listing can be found in Appendix 8.3. The step response produced by the DSP implementation is shown in Fig. 8.17; compare with Fig. 8.14.

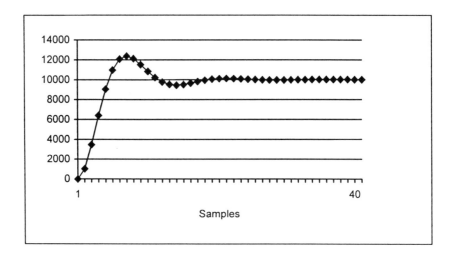

Fig 8.17: Step response from DSP implementation of Example 8.8

This is an example where a block floating point approach could be used to improve the resolution, with very little speed penalty. The multiply-accumulate operations involving r could use coefficients multiplied by 2^3, and the operations on c scaled by 2^0.

APPENDIX 8.1   TABLE OF Z-TRANSFORMS AND S-TRANSFORMS

f(t)	F(s)	f(nT)	F(z)
$\delta(t)$	1	1, n = 0 0, n ≠ 0	1
1	$1/s$	1	$1/(1 - z^{-1})$
t	$1/s^2$	nT	$Tz^{-1}/(1 - z^{-1})^2$
e^{-at}	$1/(s + a)$	e^{-anT}	$1/(1 - e^{-aT} z^{-1})$
te^{-at}	$1/(s + a)^2$	nTe^{-anT}	$Te^{-aT} z^{-1}/(1 - e^{-aT} z^{-1})^2$
$1 - e^{-at}$	$\dfrac{a}{s(s + a)}$	$1 - e^{-anT}$	$\dfrac{\left(1 - e^{-aT}\right) z^{-1}}{\left(1 - z^{-1}\right)\left(1 - e^{-aT} z^{-1}\right)}$
sin at	$\dfrac{a}{s^2 + a^2}$	sin anT	$\dfrac{\left(\sin aT\right) z^{-1}}{1 - \left(2\cos aT\right) z^{-1} + z^{-2}}$
cos at	$\dfrac{s}{s^2 + a^2}$	cos anT	$\dfrac{1 - \left(\cos aT\right) z^{-1}}{1 - \left(2\cos aT\right) z^{-1} + z^{-2}}$

APPENDIX 8.2 LMS ADAPTIVE FILTER LISTING FOR THE TMS320C25

```
                    ;LMS adaptive filter for TMS320C25/C50
                    include 'h32025.inc'

N                   equ 32      ;filter length
xin                 equ 0       ;x input port
din                 equ 1       ;d input port
yout                equ 0       ;y output port

                    ramaddr 200h
Wn                  ramspace N  ;W array
dm_convolve         ramspace 1  ;convolution start
pm_convolve         equ dm_convolve+0ff00h-200h          ;on-chip start

                    ramaddr 300h

x0                  ramspace N-1  ;input array
xn                  ramspace 1    ;last
xs                  ramspace 1    ;spare
y                   ramspace 1    ;output
d                   ramspace 1    ;reference
e                   ramspace 1    ;error
                    ;load convolution to on-chip RAM
                    larp ar1
                    lrlk ar1,dm_convolve
                    rptk on_chip_end-convolve
                    blkp convolve,*+
                    ldpk h.x0    ;page pointer
                    ; START ALGORITHM COMPUTATION
                    ;compute W = W + u*e*x
                    ;       = W + u*(d*x - y*x)
                    cnfp        ;W to program memory
                    larp ar1
                    b pm_convolve    ;to on-chip convolve
                    ;convolution below executed on chip
convolve            lrlk ar1,xn
                    zac
                    spm pright6 ;don't overflow
                    rptk N-1
                    macd 0ff00h,*-   ;convolve+dmov
                    apac        ;last +
                    sach l.y,shift2    ;new y
                    out l.y,yout
                    b lms
on_chip_end
;
```

```
lms             cnfd         ;W to data memory
                spm pshift0     ;no p shift
                lac l.d,shift7      ;get d(k)
                sub l.y,shift7     ;acc=d(k)-y(k)
                sach l.e,noshift;u*e
                lt l.e             ;t=u*e
                lrlk ar1,xs       ;ar1-> xn(after dmov)
                lrlk ar2,Wn      ;ar2-> Wn
                lark ar7,N-1     ;counter
lmsloop         mpy *-,ar2       ;p=u*e(k)*x(k)
                pac
                add *,shift15    ;new W
                sach *+,shift1,ar7 ;update W
                banz lmsloop,*-,ar1
                in l.x0,xin       ;read x
                in l.d,din        ;read d
                cnfp
lmsfin          b pm_convolve
```

APPENDIX 8.3 DIGITAL CONTROL LISTING FOR THE TMS320C25

```
;Control example for TMS320C25/C50

          include 'h32025.inc'

rin                  equ 0                        ;r input port
cout                 equ 0                        ;c output port

          ramaddr 300h
;define space for variables
cnm1                 ramspace 1                   ;c(n-1)
cnm2                 ramspace 1                   ;c(n-2)
cnm3                 ramspace 1                   ;c(n-3)
rnm1                 ramspace 1                   ;r(n-1)
rnm2                 ramspace 1                   ;r(n-2)
rnm3                 ramspace 1                   ;r(n-3)
;define constants
a2val                equ -26617 ;32768*-.8123
a3val                equ 7720    ;32768*.2356
b1val                equ 1478                     ;32768*.0451
b2val                equ 186                      ;32768*.00567
b3val                equ -1130                    ;32768*-.0345
kval                 equ 18638                    ;32768 * 2.275/4
;make space in ram for constant copies
a2                   ramspace 1
a3                   ramspace 1
b1                   ramspace 1
b2                   ramspace 1
b3                   ramspace 1
k                    ramspace 1
;
;load constants to ram
          ldpk h.cnm1                             ;data page ptr
          lalk a2val,noshift
          sacl l.a2,noshift
          lalk a3val,noshift
          sacl l.a3,noshift
          lalk b1val,noshift
          sacl l.b1,noshift
          lalk b2val,noshift
          sacl l.b2,noshift
          lalk b3val,noshift
          sacl l.b3,noshift
          lalk kval,noshift
          sacl l.k,noshift
;
loop                 lt l.rnm3
```

```
mpy l.b3  ;P=b3*r(n-3)
lack 0                              ;A=0
ltd l.rnm2 ;A=b3*r(n-3)
mpy l.b2
ltd l.rnm1 ;+b2*r(n-2)
mpy l.b1
lta l.cnm3                          ;+b1*r(n-1)
mpy l.a3
ltd l.cnm2                          ;+a3*c(n-3)
mpy l.a2
ltd l.cnm1                          ;+a2*c(n-2)
add l.cnm1,shift15 ;+1*c(n-1)
sach l.cnm1,shift1 ;store c(s)
lt l.cnm1
mpy l.k
pac                                 ;c(n)/4
sach l.cnm1,shift3 ; c(n)
out l.cnm1,cout
in l.rnm1,rin                       ;read r(n)
b loop
```

References

[1]   Analog Devices *ADSP2100 Family User's Manual*
[2]   Analog Devices *ADSP21020 User's Manual*
[3]   AT&T *WE DSP32C Digital Signal Processor*
[4]   University of Birmingham TMS320C25 assembler/simulator, Internet
       www.@eee.bham.ac.uk/dspbook
[5]   C.S. Burrus and T.W. Parks, *DFT/FFT and Convolution Algorithms:
       Theory and Implementation*, Wiley 1985.
[6]   C.S. Burrus et al, *Computer based exercises for signal processing
       using Matlab*, Prentice-Hall, 1994.
[7]   V. Cappellini et al, *Digital filters and their applications*, Academic
       Press, 1978.
[8]   D.G Childers et al, The *Cepstrum : A guide to processing*, Proc.
       IEEE, Vol. 65, No. 10, October 1977, p. 1428-1443.
[9]   R.E. Crochiere and L R Rabiner, *Multirate digital signal processing*,
       Prentice-Hall, 1989.
[10]   R.E. Crochiere and L.R. Rabiner, A program for multistage decimation,
        interpolation and narrow band filtering, in *Programs for digital signal
        processing*, IEEE Press, Wiley, 1979
[11]   D.J. DeFatta et al, *Digital signal processing*, Wiley, 1988.
[12]   F.J. Harris, *On the use of windows for harmonic analysis with the Discrete
        Fourier Transform*, Proc. IEEE, Vol 66, No 1, January 1978, pp 51-83.
[13]   S. Haykin, *Adaptive filter theory*, 2nd ed., Prentice-Hall, 1991.
[14]   IEEE, *IEEE754 Standard for Binary Floating Point Arithmetic*, IEEE New
        York 1985
[15]   P. Katz, *Digital control using microprocessors*, Prentice-Hall, 1994.
[16]   R. Kuc, *Introduction to digital signal processing*, McGraw-Hill, 1988
[17]   J. Makhoul, *Linear prediction: A tutorial review*, Proc. IEEE, Vol. 63,
        April 1975, p.561-580.
[18]   Mathsworks Inc.,*Signal processing TOOLBOX for use with MATLAB,
        July 1992, .
[19]   Motorola, *DSP56000 Digital Signal Processor Technical Summary*
[20]   Motorola, *DSP96002 IEEE Floating Point Dual-Port Processor User's
        Manual,*
[21]   K Ogata, *Solving control engineering problems with Matlab*, Prentice-
        Hall, 1994

[22]   K. Ogata, *Discrete-time control systems*, 2nd ed., Prentice-Hall, 1995.

[23]   A V Oppenheim and R W Schafer, *Digital signal processing*, Prentice-Hall, 1975.

[24]   A.V. Oppenheim and R.W. Schafer, *Discrete-time Signal Processing*, Prentice-Hall,1989

[25]   T.W Parks and C.S. Burrus, *Digital filter design*, Wiley, 1987

[26]   A. Peled and B. Liu, *Digital signal processing*, Wiley, 1976, p. 86.

[27]   J.G. Proakis and D G Manolakis, *Digital signal processing*, 2nd edition, Maxwell-Macmillan, 1992.

[28]   R.A. Roberts and C.T. Mullis, *Digital signal processing*, Addison-Wesley, 1987.

[29]   Texas Instruments, *TMS320CIIx Second-Generation User's Guide*

[30]   Texas Instruments, *TMS320C5x User's Guide*

[31]   Texas Instruments, *TMS320C3x User's Guide*

[32]   J.R. Treicher, C.R. Johnson & M.G. Larimore, *Theory and design of adaptive filters*, John Wiley & sons, New York 1987

[33]   A.W.H. Van der Enden, *Discrete-time signal processing*, N A M Verhoeckx, Prentice-Hall, 1989.

[34]   M.E Van Valkenburg, *Analog filter design*, Holt-Saunders, 1982.

Index

SIGNAL PROCESSING IN ELECTRONIC COMMUNICATIONS
MICHAEL J. CHAPMAN, DAVID P. GOODALL, and NIGEL C. STEELE, School of
Mathematics and Information Sciences, University of Coventry

ISBN 1-898563-30-6 288 pages 1997

This text for advanced undergraduates reading electrical engineering, applied
mathematics, and branches of computer science involved with signal processing (speech
synthesis, computer vision and robotics). Serves also as a reference source in academia
and industry.

Signal processing is an important aspect of electronic communications in its role of
transmitting information, and the mathematical language of its expression is developed
here in an interesting and informative way, imparting confidence to the reader.

*Contents: Sig*nal and linear system fundamentals; System responses; Fourier methods;
Analogue filters; Discrete-time signals and systems; Discrete-time system responses;
Discrete-time Fourier analysis; The design of digital filters; Aspects of speech
processing; Appendices: The complex exponential; Linear predictive coding
algorithms; Answers.

DIGITAL SIGNAL PROCESSING: Software Solutions and Applications
J.M. BLACKLEDGE and M.J. TURNER, Department of Mathematical Sciences,
Faculty of Computing Science and Engineering, De Montfort University, Leicester

ISBN: 1-898563-48-9 *ca*. 200 pages 1998

This text is for advanced undergraduates and postgraduates reading electronic
engineering, computer science and/or applied mathematics. Complete with CD-ROM it
delivers the necessary mathematical and computational background and some of the
processing techniques used for Digital Signal Processing (DSP). The book's appeal lies
in its emphasis on software solutions for which source code is provided.

Contents: PART I: MATHEMATICAL BACKGROUND - Fourier series and Fourier
integrals; Convolution integrals; Analytical signals and the Hilbert transform; The
sampling theorem; PART II: COMPUTATIONAL BACKGROUND - Sampling and
aliasing; The convolution sum; The discrete Fourier transform; The fast Fourier
transform; Computing with FFT's; Leakage and windowing; Digital filters; The FIR
and IIR filter; PART III: PROCESSING TECHNIQUES - Inverse filters; The Wiener
filter; Constrained deconvolution; The matched filter; Bayesian estimation; Maximum
entropy filters; Non-stationary deconvolution; Super resolution techniques; Statistical
filters; Singular value decomposition; The Kalman filter; Dynamic programming
techniques; Fractal analysis of statistically self-affine signals; Wavelets.

OBJECT ANALYSIS BY IMAGE PROCESSING AND PATTERN RECOGNITION

DAISHENG LUO, Institute of Biomedical and Life Science, University of Glasgow

ISBN: 1-898563-52-7   ca.250 pages   1998

This book delivers a course module for advanced undergraduates, post-graduates and researchers reading in departments of electronics, computing science, medical imaging, or wherever the study of identification and classification of objects by electronics-driven image processing and pattern recognition is relevant. There is a need for an up-to-date book in terms of the clear focus of this authors' text in this fast moving subject area.

Contents: Image processing and pattern recognition; Orientation analysis; Object detection; Arrangement analysis; Shape analysis; Conclusions; Analysis of roundness/sharpness.

IMAGING AND DIGITAL IMAGE PROCESSING

J.M. BLACKLEDGE, Department of Mathematical Sciences, Faculty of Computing Science and Engineering, De Montfort University, Leicester

ISBN: 1-898563-49-7   ca. 350 pages with CD-ROM   2000

This book delivers an authoritative account of the theory and mathematical methods, algorithms and applications of imaging systems and the digital techniques for processing the data provided by these systems. Written for advanced undergraduates and post-graduates reading computer sciences, electronics, physics and applied mathematics, it will also serve as a reference source for professional researchers in the industrial and academic sectors. The text includes a CD-ROM which provides a small object library of image processing objects written in C/C++ and a prototype Graphical User Interface. This will allow readers to test out some of the image processing algorithms discussed in the text and to build an image processing system from scratch, tailored to suit their own interest and research portfolio.

Contents: Introduction; The Fourier transform, convolution and the sampling theorem; The discrete Fourier transform; Field equations and wave equations; Green functions and scattering theory; The incoherent imaging equation; Projection tomography; Some applications in astronomy and remote sensing; Imaging of layered materials; Diffraction tomography; Coherent optical imaging; Synthetic aperture imaging systems; Deconvolution; Extraction of information from noise; Image enhancement; Image segmentation and pattern recognition; Image compression; Conclusions and future research.

OBJECT-ORIENTED TECHNOLOGY AND COMPUTING SYSTEMS RE-ENGINEERING

H. ZEDAN and A. CAU, Software Technical Research Laboratory, De Montfort University, Leicester

ISBN: 1-898563-56-X   *ca.* 200 pages   1998

This book delivers, for advanced study and R&D in computing science and electronics engineering, the latest developments in object technology and their impact in computing systems re-engineering; Object-oriented programming is here shown to provide support for constructing large scale systems, cheaply built and with re-usable components, adaptable to changing requirements by efficient and cost-effective techniques.

Contents: Formal Methods in OO: Towards an object-oriented design methodology for hybrid systems; Fair objects; System Re-engineering: Re-engineering requirements specifications for reuse: A synthesis of 3 years industrial experience; Pre-processing COBOL programs for reverse engineering in a software maintenance tool; Re-engineering procedural software to object-oriented software using design transformations and resource usage matrix; Using OO design to enhance procedural software; Devising coexistence strategies for objects with legacy systems; OO Languages: Systems of systems as communicating structures; Applications of OO: Object-oriented development of x-ray spectrometer software; Object-oriented model for expert systems implementation; Design Patterns and CORBA: Design patterns and their role in formal object-oriented development; Suitability of CORBA as a heterogeneous distributed platform.

IMAGE PROCESSING RESEARCH: Mathematical Methods, Algorithms & Applications

J.M. BLACKLEDGE, Department of Mathematical Sciences, Faculty of Computing Science and Engineering, De Montfort University, Leicester

ISBN: 1-898563-61-6   *ca.* 300 pages   1999

International specialists address state-of-the-art research in the modelling, processing and analysis of digital images. They focus on applications in science, civil and defence industries, many associated with the application of advanced imaging systems. The book includes presentations of hardware/software and imaging systems developed by a range of industries and research institutes.

The book stems from the 1998 Image Processing Conference held under the auspices of the Institute of Mathematics and Applications in De Montfort University, Leicester.

It contains contributions from scientists from Belgium, France, Germany, Poland, Spain, UK and USA, classified into:

               Part I   - MATHEMATICAL MODELLING
               Part II - INVERSE PROBLEMS (image restoration and reconstruction)
               Part III - IMAGE COMPRESSION
               Part IV - FRACTALS AND WAVELETS IN IMAGE PROCESSING
               Part V - PATTERN RECOGNITION
               Part VI - IMAGE UNDERSTANDING.

CALCULUS: Introduction to Theory and Applications in Physical and Life Science
R.M. JOHNSON, Department of Mathematics and Statistics, University of Paisley
ISBN: 1-898563-06-3   336 pages   1995

This lucid and balanced text for first year undergraduates (UK) conveys the clear understanding of the fundamentals and applications of calculus, as a prelude to studying more advanced functions. Short and fundamental diagnostic exercises at chapter ends testing comprehension, before moving to new material.

Contents: Prerequisites from algebra, geometry and trigonometry; Limits and differentiation; Differentiation of products and quotients; Higher-order derivatives; Integration; Definite integrals; Stationary points and points of inflexion; Applications of the function of a function rule; The exponential, logarithmic and hyperbolic functions; Methods of integration; Further applications of integration; Approximate integration; Infinite series; Differential equations.

LINEAR DIFFERENTIAL AND DIFFERENCE EQUATIONS:
A systems approach for mathematicians and engineers
R.M. JOHNSON, Department of Mathematics and Statistics, University of Paisley
ISBN: 1-898563-12-8   200 pages   1996

This text for advanced undergraduates and graduates reading applied mathematics, electrical, mechanical, or control engineering employs block diagram notation to highlight comparable features of linear differential and difference equations, a unique feature found in no other book. The treatment of transform theory (Laplace transforms and z-transforms) encourages readers to think in terms of transfer functions, i.e., algebra rather than calculus. This contrives short-cuts whereby steady-state and transient solutions are determined from simple operations on the transfer functions.

"Should find wide application by undergraduate students in engineering and computer science ... the author is to be congratulated on the importance that he attaches to conveying the parallelism of continuous and discrete systems" - *Institute of Electrical Engineers (IEE) Proceedings*

ORDINARY DIFFERENTIAL EQUATIONS: A concise course in applied mathematics, physics, engineering
W.S. WEIGLHOFER and K.A. LINDSAY, Department of Mathematics, University of Glasgow
ISBN: 1-898563-57-8   144 pages   1999

This advanced undergraduate text provides the basis for a semester/module of 20-25 lectures for students of applied mathematics and the applied sciences. It approaches the study of ordinary differential equations from two directions: (1) in the more traditional setting, first-order differential equations and their solutions; (2) followed by a practical and more modern approach to mathematical modelling. Thereby it emphasizes how various types of differential equations arise in simple modelling scenarios in applied mathematics and physics, the engineering and biological sciences.

Contents: Differential equations of the first order: Revision; Modelling applications of the first order; Linear differential equations of the second order: Introduction; Homogeneous equations with constant coefficient; Solution of the inhomogeneous equation; Damped oscillatory motion; Forced oscillations; Other solution methods for second order equations; Laplace transform; Euler's Differential Equation; Linear equations: Higher order initial value problems; Systems of first order linear equations; Boundary value problems; optimization and calculus variations: optimization; calculus of variations.

NOTES

NOTES

NOTES

NOTES

NOTES